Springer Theses

Recognizing Outstanding Ph.D. Research

For further volumes:
http://www.springer.com/series/8790

Aims and Scope

The series "Springer Theses" brings together a selection of the very best Ph.D. theses from around the world and across the physical sciences. Nominated and endorsed by two recognized specialists, each published volume has been selected for its scientific excellence and the high impact of its contents for the pertinent field of research. For greater accessibility to non-specialists, the published versions include an extended introduction, as well as a foreword by the student's supervisor explaining the special relevance of the work for the field. As a whole, the series will provide a valuable resource both for newcomers to the research fields described, and for other scientists seeking detailed background information on special questions. Finally, it provides an accredited documentation of the valuable contributions made by today's younger generation of scientists.

Theses are accepted into the series by invited nomination only and must fulfill all of the following criteria

- They must be written in good English
- The topic should fall within the confines of Chemistry, Physics and related interdisciplinary fields such as Materials, Nanoscience, Chemical Engineering, Complex Systems and Biophysics.
- The work reported in the thesis must represent a significant scientific advance.
- If the thesis includes previously published material, permission to reproduce this must be gained from the respective copyright holder.
- They must have been examined and passed during the 12 months prior to nomination.
- Each thesis should include a foreword by the supervisor outlining the significance of its content.
- The theses should have a clearly defined structure including an introduction accessible to scientists not expert in that particular field.

Beth Shaw

Active Tectonics of the Hellenic Subduction Zone

Springer

Author
Dr Beth Shaw
St John's College
University of Cambridge
Cambridge
UK
e-mail: bs370@cam.ac.uk

Supervisor
Prof. James Jackson
Bullard Laboratories
Department of Earth Sciences
University of Cambridge
Cambridge
UK
e-mail: jaj2@cam.ac.uk

ISSN 2190-5053
e-ISSN 2190-5061
ISBN 978-3-642-20803-4
e-ISBN 978-3-642-20804-1
DOI 10.1007/978-3-642-20804-1
Springer Heidelberg Dordrecht London New York

Printed on acid-free paper

Springer is part of Springer Science+Business Media (www.springer.com)

The object of my mission had sufficient interest in itself to excite a feeling of enthusiasm, from an anticipation of future utility, from the opportunity afforded me of studying some of the grander features of nature, and from the varied interests connected with the examination and exploration of a new field.

Thomas A.B. Spratt

Travels and Researches in Crete, 1865

Publications resulting from this thesis

The material described in Chap. 2 has been published as: Shaw et al. 2008, Eastern Mediterranean tectonics and tsunami hazard inferred from the AD 365 earthquake. *Nature Geoscience* 1, 268–276.

The material described in Chap. 3 has been published as: Shaw and Jackson 2010, Earthquake mechanisms and active tectonics of the Hellenic subduction zone. *Geophysical Journal International* doi:10.1111/j.1365-246X.2010.04551.x

The material described in Chap. 4 has been published as: Shaw et al. 2010, Radiometric dates of uplifted marine fauna in Greece: Implications for the interpretation of recent earthquake and tectonic histories using lithophagid dates. *Earth and Planetary Science Letters* 297, 395–404.

Supervisor's Foreword

Beth Shaw's thesis illustrates very well the progress, and unexpected benefits, of curiosity-driven science at its best, ranging widely over diverse fields of endeavour, from earthquake seismology, to coastal geomorphology, elastic dislocation modelling and the implications of marine ecology and biogeochemistry for radiocarbon dating. It is through making connections across such scientific disciplines that our view of the planet we live on can change profoundly. It is not an easy course of research to pursue, and requires determination, and some confidence, to follow clues without even knowing what they are clues to. This approach is not for the specialist who is comfortable only within the well-defined boundaries of a particular professional discipline, yet it is manifestly the way to find out about a planet that is constantly carrying out experiments without the courtesy of informing us first what those experiments are for.

The thesis began with a forensic investigation of a famous catastrophe of the Byzantine world, in which a tsunami destroyed Alexandria, the Nile delta, and other coastal settlements in Sicily and the Adriatic in AD 365. The prime suspect for causing the tsunami was an earthquake in the Hellenic subduction zone, but it was also known that the 40 mm/year convergence between Crete and Libya was mostly accommodated without earthquakes, which can only account for about 10% of the necessary motion. There is nothing particularly unusual about slip without earthquakes, even in other subduction zones, but it is generally thought to be a characteristic of the slip surface itself, which is either dominated by friction (and generates earthquakes) or not. If the Hellenic subduction zone generated occasional earthquakes of magnitude 8 or greater it could evidently behave in both ways, which was a puzzle. Another element in the mystery was the remarkable observation, first recorded by Captain Spratt RN in 1865, that the entire SW corner of Crete had been uplifted by up to 10 m since Roman times, but by what mechanism? Following these clues and resolving these puzzles was the first element of Beth's thesis and, like any reviewer reluctant to give away the plot of a detective story, I will not reveal the answer in this preface. It is sufficient to point out that its discovery involved an unusually broad spectrum of observations, arguments and techniques, including earthquake seismology, GPS, radiocarbon dating, dislocation and tsunami modelling and field observations of geomorphology.

Solving this first mystery of course revealed others. It showed how the convergence between Africa and Crete was accommodated near Crete, and how it was able to generate tsunamis. But what was the tsunami potential of the rest of the arc between the Ionian islands and Rhodes? How do the variations in earthquake mechanisms relate to the motions revealed by GPS measurements? How does the active faulting in the subduction zone connect with that in central Greece, and what happens at the eastern and western ends of the arc? These questions were answered in the second part of the thesis, which involved a re-evaluation of the earthquakes in the entire arc, a major effort in observational seismology, and a close examination of the GPS measurements over the last 10 years. Several important new insights were revealed by this investigation, which represented a step change in clarifying the patterns of earthquakes and what they were telling us.

The final substantial element of the thesis involved an entirely different focus. Having established that SW Crete was indeed uplifted in a single event in AD 365, an outstanding puzzle was why some uplifted marine shells, which must have been killed in that event, gave the 'wrong' age when carbon-dated; usually too old by up to 1,000 years. This was more than a curiosity because the shells concerned are the best and most commonly preserved organisms in uplifted coastlines around much of the Mediterranean and Aegean, and are often used to estimate tectonic uplift rates. The AD 365 uplift event allowed Beth to investigate this effect and discover that it is related to the ecology and mode of life of the shells themselves, which incorporate old (radiocarbon-dead) carbon from host limestone into their shells during growth. Quantification of this effect reveals the limitations of these shells in understanding uplift rates in much wider tectonic contexts than Crete.

Other questions and puzzles have been revealed by this work and remain to be pursued. But the work reported here was substantially completed in 3 years, which is the intended duration of a Ph.D. study, and so a halt had to be called. But it is not a halt to the work's influence: all of it has now been published in professional journals, and I expect it will have an important effect on tectonic studies in the Mediterranean and other subduction zones for some time. No one could ask more of a thesis.

Cambridge, 19 January 2011 James Jackson

Acknowledgements

Sincere thanks are due to my supervisor, James Jackson; without him this thesis could not have been written. I am extremely fortunate in having been able to discuss my work with Philip England, Dan McKenzie and Tom Higham, whose comments and advice have dramatically improved this thesis.

The tsunami modelling described here was carried out by Matthew Piggott and colleagues at Imperial College, London. Michael Floyd and Jean-Matthieu Nocquet supplied the GPS measurements that were used extensively in this thesis. Nic Ambraseys is an absolute mine of information concerning the historical record of the Mediterranean, and it was a great pleasure to discuss ancient descriptions of earthquakes with him. Microearthquake data was generously provided by Denis Hatzfeld and Thomas Meier.

Special thanks are also due to Martin Brasier, Norman Charnley, Jill Darrell, Owen Green, Liz Harper, Gideon Henderson, Mathew Lowe, Roberto Portela-Miguez, Richard Preece, Gareth Roberts, Brian Rosen, Alastair Sloan, Paul Taylor, Alex Thomas, Kathie Way, Tim Wright, and the staff of the NERC Radiocarbon centres in Oxford and East Kilbride.

I owe an enormous debt of gratitude to my friends in St John's College and at Bullard Laboratories. My family have been a continual source of support, and have always encouraged me to be curious about the world around me. Finally, my sincere thanks go to my husband, Richard, to whom I dedicate this thesis. Without his constant cheerfulness, encouragement and calm in the face of innumerable minor Ph.D. disasters, this thesis would probably not exist.

Contents

Chapter 1
Introduction

The Hellenic subduction zone plays a key role in the active tectonics of the Mediterranean, yet its tsunamigenic potential and details of its kinematics remain poorly understood. The subduction zone appears to be largely uncoupled, having accommodated just 10% of the Nubia-Aegean convergence in earthquakes in the last 100 years. However, the historical record contains evidence of devastating earthquakes and tsunamis, raising the possibility that the subduction zone interface fails in large magnitude earthquakes with a repeat time that is long compared with the period of observation. The focus of this thesis is an investigation of the kinematics of the region, beginning with a forensic investigation of the AD 365 earthquake, based on evidence found on the island of Crete.

Mesozoic Nubian lithosphere subducts to the North beneath overriding Aegean material in the Hellenic subduction zone (Fig. 1.1). Nubia is moving north towards Eurasia at just 5–10 mm/year, however due to the south-west motion of the Aegean with respect to Eurasia, 35 mm/year must be accommodated across the Hellenic subduction zone [1, 2].

The subducting slab appears as a relatively fast velocity anomaly to depths of at least 600 km [3–5] but is asesimic below 180 km depth [6]. At depth, the slab is almost semicircular in shape because of the highly arcuate trench system, but steepens dramatically from west to east [4, 7–11].

The oceanic Moho of the downgoing African plate can be identified using receiver functions at depths of between 40 and 69 km beneath Crete [11–14]. It is possible to follow this interface to a depth of $\sim$100 km beneath Santorini [13], or to around 160 km beneath the volcanic arc generally and to 220 km beneath northern Greece [11]. Most receiver function studies recognise a low-velocity region between 20 and 50 km, which has been explained as due either to subjected sediments [12], or as serpentenised Aegean upper mantle [11, 14].

The subduction zone interface is aseismic between the surface and $\sim$15 km depth [15–17], with shallowly dipping thrust-faulting earthquakes observed at depths of 15–45 km. Such interface events reach a depth of 45 km beneath the SW corner of Crete, approximately 10 km above the receiver function estimates of African Moho

B. Shaw, *Active Tectonics of the Hellenic Subduction Zone*, Springer Theses,
DOI: 10.1007/978-3-642-20804-1_1, © Springer-Verlag Berlin Heidelberg 2012

depths in this area, consistent with the subduction zone interface coinciding with the top of the African oceanic crust.

The surface expression of the subduction zone is obscured by up to 10 km of deformed sediment. No true subduction zone trench can be recognised in the bathymetry, in that the interface fault is probably 'blind', projecting to the surface beneath the Mediterranean Ridge, a thick accretionary prism (Fig. 1.1). The prism is actively deforming and mud-volcanoes, landslides, salt domes and faults can be recognised on its surface [18–22]. The sediments of the accretionary prism contain a number of weak layers, including Messinian salt and Cretaceous shales along which sliding may occur [23, 24]. Bohnhoff et al. [25] carried out a wide-aperture seismic experiment, using a combination of OBS and onshore data. They observed material with the velocity of continental crust, which may be compacted sediments, thinning to the south of Crete to a minimum of 17 km, and observed the contact between this material and oceanic crust ~100 km south of Crete, beneath the central Mediterranean Ridge.

A clear bathymetric escarpment 3 km deep, known as the Hellenic Trench, runs from the Ionian Islands to the south-west of Crete, splitting into at least three branches south of Crete, and continuing east to Rhodes as the Ptolemy, Pliny and Strabo trenches [18, 26, 27]. The Hellenic Trench is not the surface expression of the subduction interface, but may be a backstop to the accretionary prism.

Crete lies to the north-east of the Hellenic trench, and elevated marine deposits, deeply incised river gorges and the high, recent topography in an area affected by E-W extension suggest that the island has been rapidly uplifted since at least the Middle Miocene. Le Pichon and Angelier [28] suggested that the observed uplift results from sediment underplating. In spite of the great thickness (up to 10 km) of sediment entering the subduction zone, the isotopic composition of igneous material erupted from Santorini and Milos contain almost no sedimentary signature [29, 30], suggesting that the sediments are not subducted to depths at which melting occurs. In a more recent study, Zellmer et al. [31] found that their samples from Santorini could be modelled with the addition of a very small fraction of sediment-derived partial melt to the mantle wedge (0.2–0.4%). The buoyant sediments may be partially scraped off to form the accretionary prism, or subducted to shallow depths beneath Crete until they can be subducted no further, uplifting Crete by crustal thickening.

Despite being the most seismically active region in Europe, just 10% of the expected convergence between Nubia and the southern Aegean has been accommodated by earthquakes during the instrumental record [32]. The seismic deficit could be due to aseismic slip in the subduction zone, or due to the period of observation being short compared to the repeat time of large magnitude earthquakes that accommodate the majority of the convergence. To distinguish between these two possibilities, it is necessary to examine the long and rich historical record of the region. In Chap. 2, I address this question by carrying out a forensic study of the AD 365 earthquake which occurred close to the island of Crete (Fig. 1.1). I suggest that the subduction zone interface is largely aseismic, but that the Hellenic Trench is the surface expression of a splay fault which is capable of producing earthquakes larger

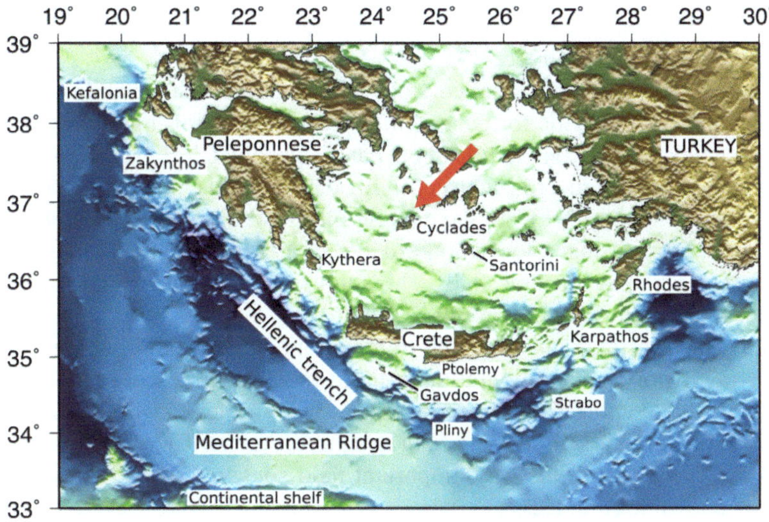

Fig. 1.1 Major topographic and bathymetric features of the Hellenic subduction zone. *Red arrow* represents velocity of the south Aegean relative to the Nubian (North African) plate

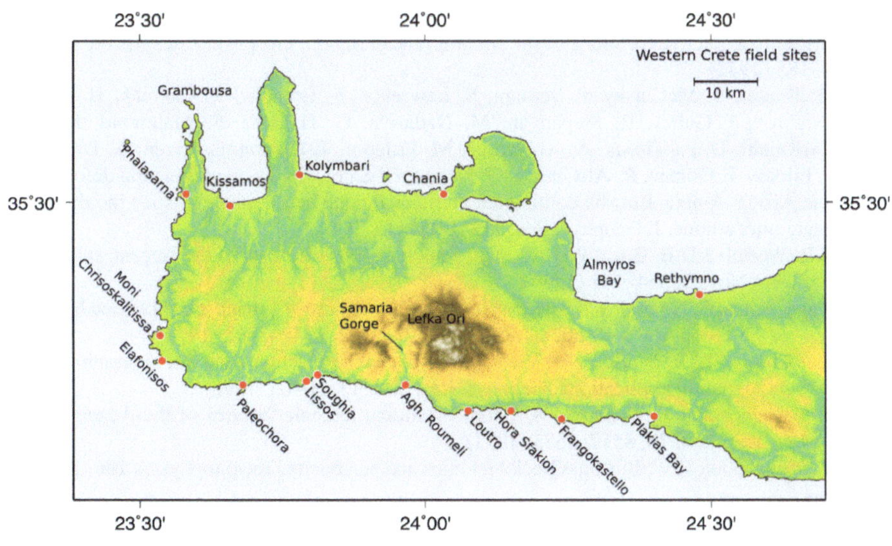

Fig. 1.2 Important field sites and towns in western Crete, referred to throughout this thesis

than magnitude 8 and tsunamis that travel throughout the Eastern Mediterranean, and estimate the approximate repeat time for earthquakes of this magnitude.

Further questions related to the kinematics of the region remain, and in Chap. 3, I use teleseismic waveform modelling to elucidate the structure of the Hellenic subduction zone, focusing in particular on the processes occurring at its ends, and suggest a connection between the Hellenic subduction zone and normal faulting and block rotation in Central Greece.

In Chap. 4, I carry out a detailed study of the utility and reliability of corals, bryozoans and lithophagids for constraining tectonic uplift using radiocarbon dating. This study arose from radiocarbon dates described in Chap. 2 which show that lithophagids may record incorrect uplift dates. In this chapter, I confirm that lithophagid dates are offset from the date of uplift and investigate possible reasons for this offset, discussing colonisation order, preservation potential, and incorporation of detrital carbonate in their shells. This study has important general implications for radiocarbon dating tectonic events.

In order to investigate the potential of topography and geomorphology for constraining the history and configuration of uplift in Crete, I present digital elevation models (DEMs) of uplifted marine terraces and river profiles constructed using kinematic GPS in Chap. 5. I discuss the long term uplift rate of Crete and the accuracy of DEMs constructed in this way. Important field sites, referred to throughout the thesis, are shown in (Fig. 1.2).

References

1. D. McKenzie, Active tectonics of the Mediterranean region. Geophys. J. R. Astron. Soc. **30**, 109–185 (1972)
2. R. Reilinger, S. McClusky, P. Vernant, S. Lawrence, S. Ergintav, R. Cakmak, H. Ozener, F. Kadirov, I. Guliev, R. Stepanyan, M. Nadariya, G. Hahubia, S. Mahmoud, K. Sakr, A. ArRajehi, D. Paradissis, A. Al-Aydrus, M. Prilepin, T. Guseva, E. Evren, A. Dmitrotsa, S.V. Filikov, F. Gomez, R. Al-Ghazzi, G. Karam, GPS constraints on continental deformation in the Africa–Arabia–Eurasia continental collisional zone and implications for the dynamics of plate interactions. J. Geophys. Res. **111**, B05411 (2006)
3. M.J.R. Wortel, S.D.B. Goes, W. Spakman, Structure and seismicity of the Aegean subduction zone. Terra Nova **2**, 554–562 (1990)
4. C. Papazachos, G. Nolet, P and S deep velocity structure of the Hellenic area obtained by robust nonlinear inversion of travel times. J. Geophys. Res. **102**, 8349–8367 (1997)
5. W. Spakman, M.J.R. Wortel, N.J. Vlaar, The Hellenic subduction zone a tomographic image and its geodynamic implications. Geophys. Res. Lett. **15**, 60–63 (1998)
6. B.C. Papazachos, P.E. Comninakis, Geophysical and tectonic features of the Aegean arc. J. Geophys. Res. Lett. **76**, 8517–8533 (1971)
7. K.C. Makropoulos, W. Burton, Greek tectonics and seismicity. Tectonophysics **106**, 275–304 (1984)
8. B.C. Papazachos, Seismicity of the Aegean and surrounding areas. Tectonophysics **178**, 287–308 (1990)
9. D. Hatzfeld, C. Martin, Intermediate depth seismicity in the Aegean defined by teleseismic data. Earth Planet. Sci. Lett. **113**, 267–275 (1992)
10. B.C. Papazachos, V.G. Karakostas, C.B. Papazachos, E.M. Scordilis, The geometry of the Wadati–Benioff zone and lithospheric kinematics in the Hellenic arc. Tectonophysics **319**, 275–300 (2000)
11. F. Sodoudi, R. Kind, D. Hatzfeld, K. Priestley, W. Hanka, K. Wylegalia, G. Stavrakakis, A. Vafidis, H.-P. Harjes, M. Bohnhoff, Lithospheric structure of the Aegean obtained from P and S receiver functions. J. Geophys. Res. **111**, B12307 (2006)
12. M. Knapmeyer, H.-P. Harjes, Imaging crustal discontinuities and the downgoing slab beneath western Crete. Geophys. J. Int. **143**, 1–21 (2000)

13. X. Li, G. Bock, A. Vafidis, R. Kind, H.-P. Harjes, W. Hanka, K. Wylegalla, van der M. Meijde, X. Yuan, Receiver function study of the Hellenic subduction zone: imaging crustal thickness variations and the oceanic Moho of the descending African lithosphere. Geophys. J. Int. **155**, 733–748 (2003)
14. B. Endrun, T. Meier, M. Bischoff, H.-P. Harjes, Lithospheric structure in the area of Crete constrained by receiver functions and dispersion analysis of Rayleigh phase velocities. Geophys. J. Int. **158**, 592–608 (2004)
15. T. Taymaz, J. Jackson, R. Westaway, Earthquake mechanisms in the Hellenic trench near Crete. Geophys. J. Int. **102**, 695–731 (1990)
16. C. Benetatos, A. Kiratzi, C. Papazachos, G. Karakaisis, Focal mechanisms of shallow and intermediate depth earthquakes along the Hellenic arc. J. Geodyn. **37**, 253–296 (2004)
17. T. Meier, M. Rische, B. Endrun, A. Vafidis, H.P. Harjes, Seismicity of the Hellenic subduction zone in the area of western and central Crete observed by temporary local seismic networks. Tectonophysics **383**, 149–169 (2004)
18. X. Le Pichon, J. Angelier, J. Aubouin, N. Lyberis, S. Monto, V. Renard, H. Got, K. Hsu, Y. Mart, J. Mascle, D. Matthews, D. Mitropoulos, P. Tsoflias, G. Chronis, From subduction to transform motion: a seabeam survey of the Hellenic trench system. Earth Planet. Sci. Lett. **44**, 441–450 (1979)
19. X. Le Pichon, N. Lyberis, J. Angelier, V. Renard, Strain distribution over the east Mediterranean ridge: a synthesis incorporating new Sea-Beam data. Tectonophysics **86**, 243–274 (1982)
20. N.H. Kenyon, R.H. Belderson, A.H. Stride, Detailed tectonic trends on the central part of the Hellenic outer ridge and in the Hellenic trench system. Geol. Soc. Lond. **10**, 335–343 (1982)
21. K.A. Kastens, A.B. Nancy, M.B. Cita, Progressive deformation of an evaporite-bearing accretionary complex: Sea-MARCI, SeaBeam and piston core observations from the Mediterranean ridge. Marine Geophys. Res. **14**, 249–298 (1992)
22. J.P. Foucher, N. Chamot-Roocke, S. Alexandry, J.M. Augustin, S. Monti, P. Pavlakis, M. Voisset, *Multibeam bathymetry and seabed reflectivity maps of the MEDRIFF corridor across the eastern Mediterranean Ridge*, Terra Cognita (Ed.), EUG (VII), London, Blackwell Scientific Publication, 278–279 (1993)
23. A. Camerlenghi, M.B. Cita, W. Hieke, T. Ricchiuto, Geological evidence for mud diapirism on the Mediterranean ridge accretionary complex. Earth Planet. Sci. Lett. **109**, 493–504 (1992)
24. J. Mascle, E. Chaumillon, An overview of Mediterranean Ridge collisional accretionary complex as deduced from multichannel seismic data. Geo-Marine Lett. **18**, 81–89 (1998)
25. M. Bohnhoff, J. Makris, D. Papanikolaou, G. Stavrakakis, Crustal investigation of the Hellenic subduction zone using wide aperture seismic data. Tectonophysics **343**, 239–262 (2001)
26. D. McKenzie, Active Tectonics of the Alpine-Himalayan belt: the Aegean Sea and surrounding regions. Geophys. J. Int. **55**, 217–254 (1978)
27. C. Kreemer, N. Chamot-Rooke, Contemporary kinematics of the southern Aegean and the Mediterranean ridge. Geophys. J. Int. **157**, 1377–1392 (2004)
28. X. Le Pichon, J. Angelier, The Aegean sea, Philosophical Transactions of the Royal Society of London. Ser. A, Math. Sci. **300**, 357–372 (1981)
29. A.C. Mann, Trace element geochemistry of high alumina basalt–andesite–dacite–rhyodacite lavas of the main volcanic series of Santorini volcano, Greece. Contrib. Mineral. Petrol. **84**, 43–57 (1983)
30. L. Briqueu, M. Javoy, J.R. Lancelot, M. Tatsumoto, Isotope geochemistry of recent magmatism in the Aegean arc: Sr, Nd, Hf and O isotopic ratios in the lavas of Milos and Santorini. Earth Planet. Sci. Lett. **80**, 41–54 (1986)
31. G. Zellmer, S. Turner, C. Hawkesworth, Timescales of destructive plate margin magmatism: new insights from Santorini, Aegean volcanic arc. Earth Planet. Sci. Lett. **174**, 265–281 (2000)
32. J. Jackson, D. McKenzie, The relationship between plate motions and seismic moment tensors, and rates of active deformation in the Mediterranean and Middle East. Geophys. J. **93**, 45–73 (1988)

Chapter 2
The AD 365 Earthquake: Large Tsunamigenic Earthquakes in the Hellenic Trench

. . .and solid earth was shaken and trembled, the sea with its rolling waves was driven back and withdrew from the land . . .Hence, many ships were stranded as if on dry land, and since many men roamed without fear in the little that remained of the waters, to gather fish and similar things with their hands, the roaring sea, resenting this forced retreat, rose in its turn; and over the boiling shoals it dashed mightily upon islands and broad stretches of the mainland, and levelled innumerable buildings in the cities and wherever else they are to be found; so that amid the mad discord of the elements the altered face of the earth revealed marvellous sights. For the great mass of waters, returning when it was least expected, killed many thousands of men by drowning; and by the swift recoil of the eddying tides a number of ships, after the swelling of the wet element subsided, were found to have been destroyed, and the lifeless bodies of shipwrecked persons lay floating on their backs or on their faces. Other great ships, driven by the mad blasts, landed on the tops of buildings (as happened at Alexandria), and some were driven almost two miles inland, like a Laconian ship which I myself saw in passing that way near the town of Motho [Mothoni/Methoni], yawning apart through long decay.

Ammianus Marcellinus, writing after AD 378.

2.1 Summary

Historical accounts describe an earthquake and tsunami on 21 July AD 365 that destroyed cities and drowned thousands of people in coastal regions from the Nile delta to modern-day Dubrovnik. The location and tectonic setting of this earthquake have been uncertain until now. In this chapter, I present evidence from radiocarbon data and field observations that western Crete was lifted above sea-level, by up to 10 m, synchronously with the AD 365 earthquake. The distribution of uplift,

B. Shaw, *Active Tectonics of the Hellenic Subduction Zone*, Springer Theses,
DOI: 10.1007/978-3-642-20804-1_2, © Springer-Verlag Berlin Heidelberg 2012

combined with observations of present-day seismicity, suggest that this earthquake occurred not on the subduction interface beneath Crete, but on a fault dipping at about 30° within the overriding plate. Calculations of tsunami propagation, carried out by Piggott and colleagues, show that the uplift of the sea floor associated with such an earthquake would have generated a damaging tsunami through much of the eastern Mediterranean. GPS measurement of the present rate of shortening between Crete and the Cyclades can be used to estimate an approximate repeat time of ~5000 years for tsunamigenic events on this single fault in western Crete. If the same process takes place along the entire Hellenic subduction zone, such events may occur approximately once every 800 years.

2.2 Introduction

The AD 365 earthquake was felt throughout the eastern Mediterranean, and its tsunami inundated coastal sites in Africa, the Adriatic, Greece and Sicily. Figure 2.1 shows sites where earthquake damage was recorded in the historical record (yellow dots), and the sites that are known to have been inundated by the tsunami (blue triangles). Rich agricultural land in the Nile delta was abandoned as a result of the flooding, and previously populous hill cities were subsequently inhabited only by hermits [1].

The cause of the AD 365 event, and of other large earthquakes and tsunamis in the Eastern Mediterranean, such as those in AD 551 [2] and AD 1303 [3], has been enigmatic. Most of the world's devastating tsunamis are thought to be generated by slip on the interfaces between plates at subduction zones. The Hellenic subduction zone is the only such major plate boundary in the region, but appears to be largely decoupled; just 10% of the total Nubia-Aegean convergence has been accommodated in earthquakes in the last 100 years (see Chap. 3 for summary).

The key to understanding the kinematics of the area lies in a paleo-shoreline that fringes the coasts of western Crete and Antikythera. This shoreline was first described by Captain Thomas Spratt, RN, in 1851 [4]. Spratt noted many 'sea marks' around the coast of western Crete, reaching a maximum elevation of 10 m above present sea-level in the south-west corner of the island. Figure 2.2 shows some examples of his sketches, with modern-day field photographs for comparison. Because the top of the marks runs through the remains of a Roman harbour at Phalasarna, which is now 6 m above sea-level, he deduced that the land must have been raised during or after the Roman era (Fig. 2.2). Pirazzoli et al. [5] showed that this shoreline had a ^{14}C age of around 2,000 years BP and attributed its uplift to an earthquake; this earthquake was subsequently linked to the AD 365 event [6–8].

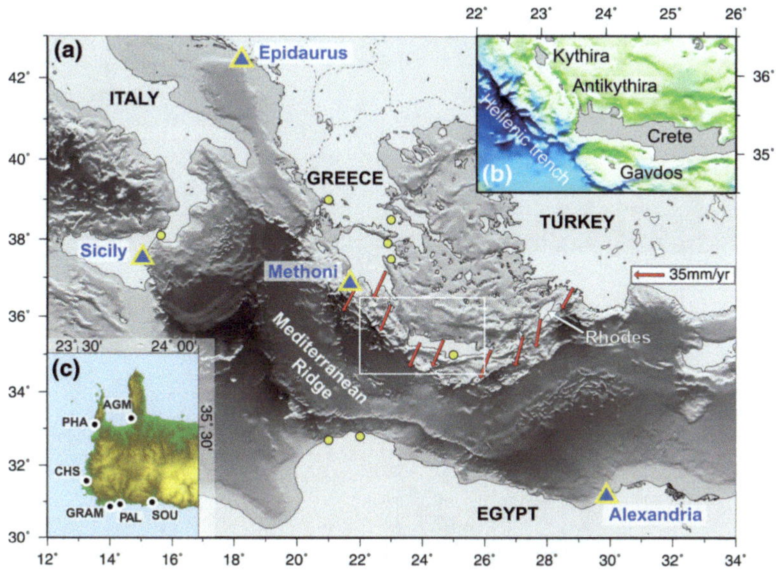

Fig. 2.1 **a** *Yellow dots* mark places where the AD 365 earthquake is recorded to have caused destruction, *blue triangles* mark places known to be affected by the associated tsunami. *Red arrows* show velocities of GPS stations near the Hellenic Trench relative to the Nubian plate from McClusky [40], McClusky et al. [41] and Reilinger et al. [42]. **b** Principal localities referred to in this chapter; note the prominent NW-SE bathymetric escarpment of the Hellenic Trench. **c** Locations of samples whose radiocarbon dates are reported in Table 2.1

2.3 Recent Surface Uplift in Western Crete and Antikythira

The starting point for this chapter is to establish whether, as suggested previously, all of the recent uplift in western Crete indeed took place in one event, close to AD 365. In what follows, the term ‘uplift’ is used to denote upward displacement of rocks relative to the geoid [9]. Because global sea-level has been stable over the interval under consideration, it is possible to take the present elevation of the paleo-shoreline as a proxy for such uplift.

A single prominent paleo-shoreline can be followed around the whole of western Crete; the algal encrustations marking this shoreline are not seen anywhere on the cliffs below this level, but similar constructions can be seen growing at sea-level today (Fig. 2.3). The elevation of the paleo-shoreline decreases from the south-western corner of Crete. Where the shoreline is highest, it is marked by a thick white algal encrustation (Fig. 2.3), but where the shoreline is lower (1–2 m above present day sea-level) it is marked by a deep erosive notch (Fig. 2.4). Previous radiocarbon dates of material from the upper levels of the paleo-shoreline were determined largely by conventional (non-AMS) methods and gave calibrated ages that lie within ~200 years of AD 365 [7, 10]. Taken together, these observations suggest that some uplift took place around AD 365, but they do not give direct evidence that the whole 10 m of

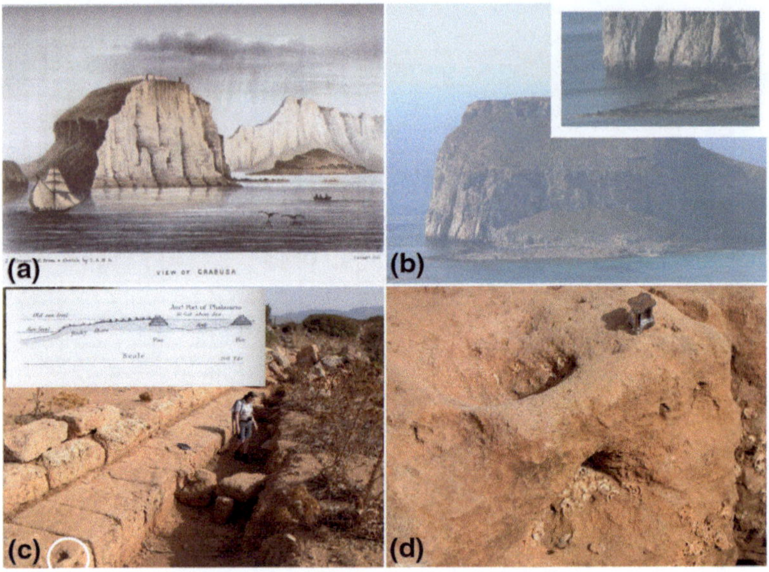

Fig. 2.2 Evidence of uplift of the western part of Crete, described by Thomas Spratt, R.N. **a** Gramvousa peninsula from an engraving by Spratt. **b** The same peninsula photographed today from 35° 35.244′N, 23° 35.682′E. The AD 365 shoreline can be recognised as a dark band around the base of the cliffs in both the engraving and the photograph (shown more clearly in the close-up *inset*). **c** The uplifted harbour at Phalasarna (35° 30.593′N, 23° 34.115′E). Encrustations on the harbour wall reveal the position of sea-level when the harbour was operational; this level is now 6 m above sea-level. The *inset* is a sketch by Spratt showing the position of the harbour walls relative to sea-level. **d** Details of an encrusted stone ring, *circled* in (**c**), that was used to tie up mooring ships

uplift took place in one step, and the attribution of the uplift to the AD 365 event was made uncertain by issues related to the calibration of the radiocarbon ages [11]. Here, I address both those sources of uncertainty with new, more accurate, radiocarbon dates on material from elevations between present sea-level and the uplifted paleo-shoreline.

Corals, bryozoans, and *Lithophaga* were collected from widely separated locations in western Crete, at elevations between 1.5 m above sea-level and the height of the paleo-shoreline (Figs. 2.5, 2.6). At all localities visited, the most recent colonisers appeared to be cheliostome bryozoans and ahermatypic corals, of which several had grown inside *Lithophaga* holes. ^{14}C dates were obtained for 13 of these corals and bryozoans. For nine of them, calibrated accelerator mass spectrometry (AMS) ^{14}C dates have 1σ ranges (typically $\sim\pm 70$ years) that include AD 365, and all but two of them include AD 365 in their 2σ age ranges (typically $\sim\pm 140$ years) (Table 2.1). All the *Lithophaga* shells gave dates older than AD 365 (see Chap. 4 for discussion).

Fig. 2.3 Clear algal encrustation marking the AD 365 shoreline. **a** At Chrisoskalitissa (35° 18.857′N, 23° 32.008′E) the top of the shoreline reaches an elevation of 8 m. **b** Close-up of the preserved algal encrustation, again from Chrisoskalitissa. **c** In the harbour at Soughia (35° 14.755′N, 23° 49.275′E) the shoreline can be recognised at 6.7 m above sea-level. **d** A similar encrustation growing at sea-level today [the photograph was taken under water in the bay to the east of the harbour at Soughia (35° 14.864′N, 23° 49.078′E)]

Before treatment, I examined the samples under an SEM to confirm the presence of primary structural fabrics (Figs. 2.5, 2.6) and selected the cleanest pieces for dating. I thoroughly cleaned all but four coral and bryozoan samples by shot blasting. Samples Phalasarna 1, Paleochora 8, Phalasarna 3 and Soughia 2 were instead cleaned ultrasonically in distilled water before being submerged in a bath of dilute (0.1 M) hydrogen peroxide until effervescence ceased.

Carbonate samples were dated using AMS ^{14}C dating by the Oxford Radiocarbon Accelerator Unit (ORAU), University of Oxford (see Chap. 4 for details). Radiocarbon determinations were calibrated using the INTCAL04 Marine Calibration dataset [12] and the OxCal computer calibration programme. AΔR value of 53 ± 43 years was used to account for the local reservoir offset (R) from the modelled world ocean [13]. This value is the average for the Eastern Mediterranean region. The calibrated data are shown in Table 2.1. The bryozoan sample Chrisoskalitissa 5 was analysed twice (17671, 17672, Table 2.1) to check reproducibility.

These new radiocarbon dates require that almost all of the uplift of western Crete took place within a few decades, i.e. within the uncertainties of dating, of AD 365. (I have no datable material from locations lower than 1.5 m above sea-level.) Although the data permit slow uplift during that interval, or uplift in a series of rapid small events, uplift of this magnitude in a single earthquake would have caused

Fig. 2.4 Erosive notch (no algal encrustation) marking the AD 365 shoreline. The top of the shoreline is relatively low (between 1 m and 2 m above present day sea-level). **a**, **b** Damiani (35° 10.49′N, 24° 24.86′E) at the eastern and western sides of the bay; **c** Plakias (35° 10.98′N, 24° 24.07′E); **d** 1 km east of Amondi Bay (35° 09.19′N, 24° 28.86′E)

the historically recorded tsunami of AD 365 (see below). Therefore either the uplift of western Crete and its surrounding seafloor took place slowly within a few decades of AD 365 and some other event caused the tsunami that destroyed Alexandria in AD 365, or more likely the two events are connected.

2.4 Fault Slip During the AD 365 Earthquake

Measurements of ground displacements after an earthquake are commonly used to place bounds on the orientation and location of the fault that slipped. In the present case, such measurements are limited to uplift, in the sense defined above, of paleoshorelines, and I use information from the distribution of earthquakes in the region as additional constraints on the location of the fault that caused the AD 365 earthquake.

The most obvious candidate fault is the subduction interface, whose orientation and location are revealed by the focal mechanisms of earthquakes striking NW-SE at 40 km depth on a shallowly-dipping (~15–20°) plane, directly below the SW corner of Crete [14], (Chap. 3, Fig. 2.7). This interface projects to the surface about 150 km SW of Crete, where the Mediterranean Ridge, a prism of sediments ~10 km thick, is deforming aseismically by folding and thrusting [15]. The interface appears to be

Fig. 2.5 **a** *Balanophyllia regia*, collected at Paleochora; **b** *Stenocyathus vermiformis*, collected at Phalasarna from within *Lithophaga* holes; **c** *Caryophyllida*, collected at Aghia Marina and imaged using SEM; **d** *Caryophyllida*, collected at Aghia Marina

aseismic beneath 40–45 km depth, in common with other subduction zones around the world [16].

Figure 2.8a–c is a cartoon showing the pattern of surface deformation during a seismic cycle in a typical subduction zone. After an earthquake, the interface is locked (a) and the overriding plate is dragged down (b). Points at the surface that are above the locked zone are dragged down, whereas points further back (above the aseismic and freely slipping deeper part of the interface) are uplifted in the flexural bulge. During an earthquake (c) the vertical motion is suddenly reversed: the region above the seismogenic part of the interface is uplifted, and points landward of this subside. Moderate-sized earthquakes on the subduction zone interface have a maximum depth of 40–45 km in this part of the Hellenic arc, and events at this depth lie directly beneath the south-west corner of Crete (Fig. 2.7, Chap. 3). If the subduction zone interface slipped between the surface and 40 km depth, Crete would lie on the landward side of the base of the seismogenic part of the interface, and would therefore be expected to subside in such an earthquake.

Figure 2.8d–e shows the calculated elastic surface deformation from an earthquake modelled as slip along the subduction zone interface between the surface and 40 km depth, using a planar dislocation in an elastic half-space [17]. The interface is assumed to dip at 15° and projects to the surface 150 km to the south-west of Crete (beneath the Mediterranean Ridge). As expected, Crete is downthrown, rather than uplifted by this earthquake, as the deepest extent of slip is seaward of the island. The

Table 2.1 ^{14}C dates of samples from western Crete (identified by Oxford Radiocarbon Accelerator Unit sample number). Age ranges in bold include AD 365

Ox ID	Organism	Location	Height (m)	Top of paleo-shore (m)	^{14}C age BP	1σ range Years	2σ range Years
16743	Coral (S)	PHA 1	3.5	6.6	1985 ± 28	AD 415–546	AD **340–604**
16745	Coral (B)	PAL 8	9.0	9.0	2019 ± 28	AD **360–507**	AD **282–565**
16991	Coral (S)	PHA 3	3.0	6.6	2032 ± 28	AD **338–484**	AD **271–551**
17669	Bryozoans (My)	CHS 1	3.3	7.9	2129 ± 27	AD **238–378**	AD **159–428**
17670	Bryozoans (My)	CHS 2	3.0	7.9	2071 ± 27	AD **297–431**	AD **235–515**
17671	Bryozoans (My)	CHS 5	2.5	7.9	2019 ± 27	AD **360–506**	AD **284–564**
17672	Bryozoans (My)	CHS 5	2.5	7.9	2024 ± 26	AD **351–500**	AD **280–556**
17673	Coral (Ca)	CHS 7	6.0	7.9	2345 ± 27	BC 22–AD 109	BC 100–AD 179
17674	Coral (Ca)	AGM	3.0	5.5	1977 ± 26	AD 423–550	AD **352–608**
17675	Bryozoans (My)	SOU NB2	2.2	6.7	2119 ± 29	AD **249–388**	AD **166–438**
17676	Bryozoans (My)	GRAM 2	4.7	9.0	1878 ± 25	AD 539–653	AD 456–675
17677	Bryozoans (My)	GRAM 4	2.0	9.0	2015 ± 28	AD **366–514**	AD **288–570**
17678	Coral (S)	PHA 2	3.4	6.6	2028 ± 27	AD **344–489**	AD **275–553**
13809	Bivalve	CHS T1	20–24	7.9	39180 ± 300	41101–40661	41851–40327
14085	Bivalve	CHS T2	20–24	7.9	37700 ± 250	40615–39981	40649–39237
16995	Bivalve	PAL T	20–24	9.0	47300 ± 550	53063–45967	53461–45621

Coral and bryozoans species were taken from or below the uplifted paleo-shoreline: *S Stenocyathus vermiformus*, *B Balanophyllia regia*, *Ca Caryophyllida*, *My Myriapora truncata*. The bivalves were taken from marine terraces at 20–24 m above sea-level, and at least 10 m above the paleo-shoreline associated with the AD 365 event. Abbreviated location names correspond to labels in inset on Fig. 2.1: *AGM* Aghia Marina, *PHA* Phalasarna, *CHS* Chrisoskalitissa, *GRAM* Gramena, *PAL* Paleochora, *SOU* Soughia. The 1σ and 2σ ages for samples 13809, 14085 and 16995 are years BC, and were obtained by comparing the radiocarbon ages with the ^{14}C record from the Cariaco Basin [45] with a 400 year subtraction made from the conventional ages to account for the marine reservoir effect

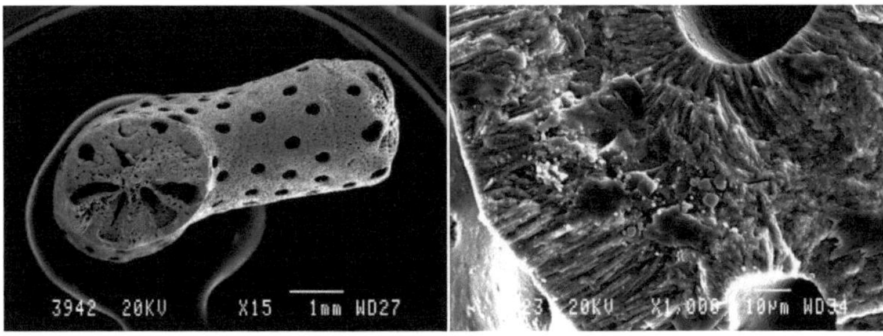

Fig. 2.6 Example of a bryozoan sample (*Myriapora truncata*) imaged using the SEM. The *first* image shows the entire organism, the *second* is a close-up of the crystalline structure. The aragonite needles radiate from the pores, and from the edge of the organism, showing that the primary crystalline fabric has been preserved

only way to force the uplift of Crete by slip along the subduction zone interface is to have coseismic slip down to depths of ~70 km, far deeper than the seismogenic thickness of the region. Slip along the subduction zone interface therefore does not seem to be a plausible mechanism for producing the AD 365 uplift.

The distribution of shallow earthquakes (Fig. 2.7) suggests that other faults may be active in the region between the Mediterranean Ridge and the coast of Crete. The most prominent surface feature in this area is the Hellenic Trench, a linear NW-SE escarpment 3,500 m deep at its lowest point, 25 km SW of Crete (Fig. 2.7). The proximity of this escarpment to the highest ground on Crete, where there is abundant geomorphological evidence of recent uplift (Chap. 5), suggests that it marks the outcrop of a major reverse fault dipping beneath Crete. The Hellenic Trench strikes at 315°, parallel to the Mediterranean Ridge and perpendicular to the slip-vectors of thrust-faulting earthquakes in this region. In order to determine possible causative faults for the AD 365 earthquake, I searched for faults that best fit the uplift data from western Crete. I restricted candidate faults for the AD 365 earthquake to faults with this strike and cropping out between the Mediterranean Ridge and the coast of Crete. This range bounds the plausible locations for reverse faults that could cause the observed uplift; a fault below the subduction interface is unlikely, a reverse fault cropping out on land would drop the south-western-most part of Crete, not raise it. I systematically searched this parameter space to find the fault that yielded the best fit (in the least-squares sense) to the observed uplift.

I modelled slip on candidate faults as planar dislocations buried within an elastic half-space [17]. I assumed that relative motion was in the dip-slip sense, as seen in modern fault-plane solutions for the region; the uplift data, in any case, do not constrain the strike-slip component of motion. I assumed that the fault slipped from the surface to a depth of 45 km, a maximum possible depth suggested by the hypocentres of micro-earthquakes within the overriding plate, which are all shallower than 45 km [18, 19], and by the maximum depths of shallowly dipping thrust faulting on the plate interface [14], Fig. 2.7 and Chap. 3. With these assumptions, the free parameters of

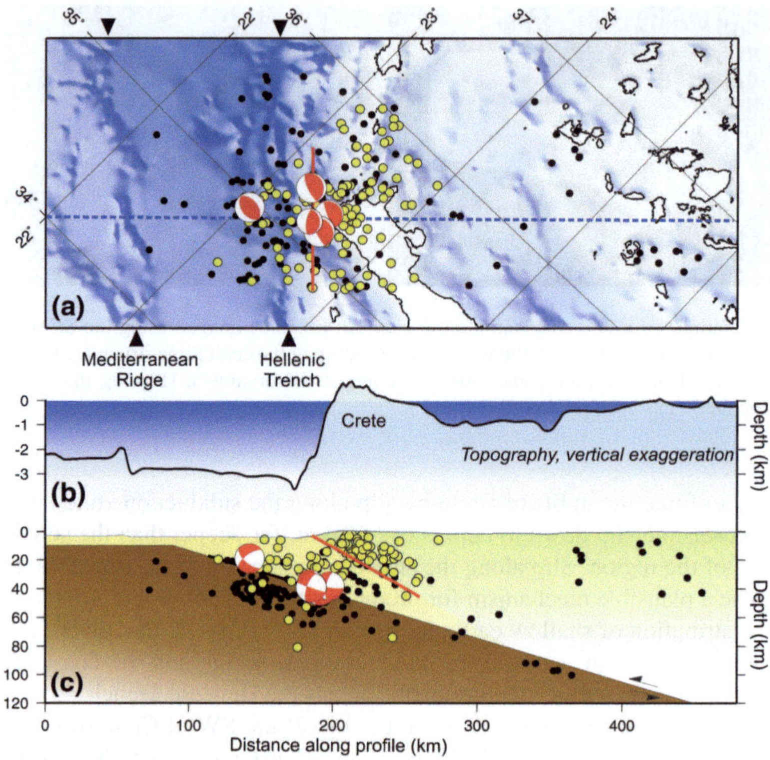

Fig. 2.7 Seismicity and topography in the area of Crete. **a** Seismicity of the region corresponding to the AD 365 earthquake. *Yellow dots* show epicentres of earthquakes determined from teleseismic data [42], *black dots* depict epicentres of microearthquakes determined from local networks [18, 43]. Fault-plane solutions for earthquakes with well-determined depths on the plate interface are from Taymaz et al. [14] and from this study. The projection to the surface of the best-fitting fault for AD 365 earthquake (Fig. 2.9) is shown by *red line*. **b** Vertically exaggerated topography showing the locations of the Mediterranean Ridge and Hellenic Trench. **c** Topography and seismicity projected onto a vertical section along *dotted line* in the centre of panel (**a**), with our interpretation of the structure. The plate interface is shown by a *black line* and focal mechanisms, the best-fitting fault (Fig. 2.9) by a *red line*. Symbols for hypocentres in (**c**) as for epicentres in (**a**). The *brown region* represents the subducting African lithosphere and the *green region* is the overriding Aegean

the candidate faults are their dips, the distance from Crete that they project to the surface, and the amount of slip. The end points of the fault are constrained by uplift of 2.7 m on Antikythira in the NW (of similar appearance and age to that on Crete, [20, 7]), and by my observation that there is no uplifted paleo-shoreline on the island of Gavdos, in the south-east, so the length of the fault was set at 100 km.

The best fit to the uplift data is obtained for a fault that crops out near the Hellenic Trench (Fig. 2.9). The fault length is 100 km, the slip is 20 ± 5 m to a depth of 45 km, the dip is $30 \pm 5°$, and the RMS misfit to the observations is 0.8 m (Fig. 2.9). If the causative fault is forced to have a dip as shallow ($\sim 15°$) as that of the plate interface,

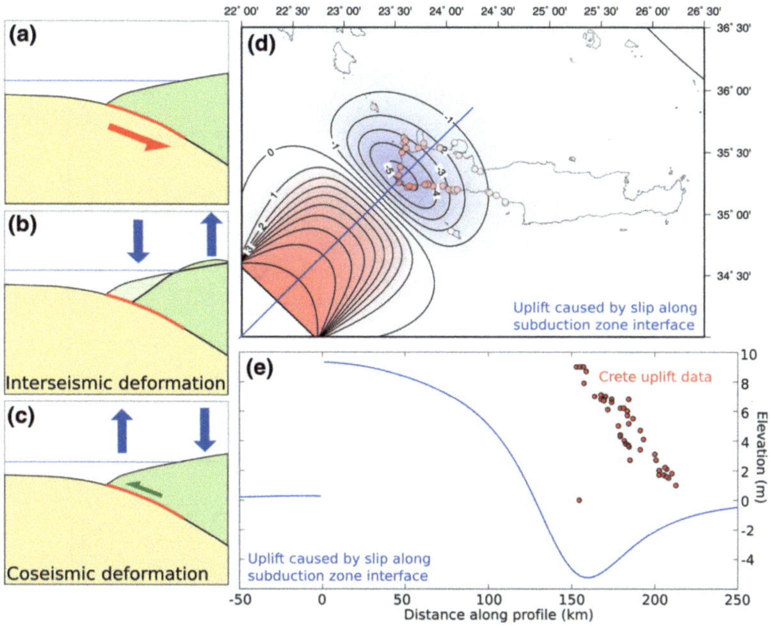

Fig. 2.8 **a–c** Cartoons showing the pattern of surface deformation during a seismic cycle in a typical subduction zone, modified from a cartoon by James Jackson. After an earthquake, the interface is locked (**a**) and the overriding plate is dragged down (**b**). Points seaward of the 'hinge' (which lies above the base of the locked zone) are dragged down, whereas points further back (above the aseismic and freely slipping part of the interface) are uplifted in the flexural bulge. During an earthquake (**c**) the vertical motion is suddenly reversed: the region above the seismogenic part of the interface are uplifted, and points landward of the 'hinge' subside. **d** Pattern of uplift expected if slip occurs along the subduction zone interface, between the surface and 45 km depth. Contours are plotted every metre, and the surface is shaded according to uplift. The island of Crete should subside, rather than being uplifted. The *pink dots* show the elevation of the AD 365 shoreline around the coast of Western Crete. The *blue line* shows the surface trace of the cross-section. **e** Cross-section of the surface deformation from the interface-slip model in (**d**), plotted as elevation (m) against distance along the profile. The AD 365 shoreline data are plotted as *red dots*

but the distance from Crete and the slip are allowed to vary, then the best-fitting solution, with an RMS misfit of 0.9 m, yields a fault that crops out only 70–80 km SW of Crete, where there is no obvious topographic expression of faulting. Any fault dipping at $\sim 15°$ requires slip of 45 ± 10 m to match the observed uplift, which is greater than has been recorded for any earthquake. The trade-offs between parameters are illustrated in Fig. 2.10. The red dot in each of the plots represents the global minimum (i.e. the solution that best fits the measured uplift data). Contours show increasing misfit (in metres) between the model and the data. If the fault is constrained to have a shallow dip, more slip must occur to produce the observed uplift (Fig. 2.10b). The blue dot shows the best-fitting solution if the dip of the fault is constrained to be 15°.

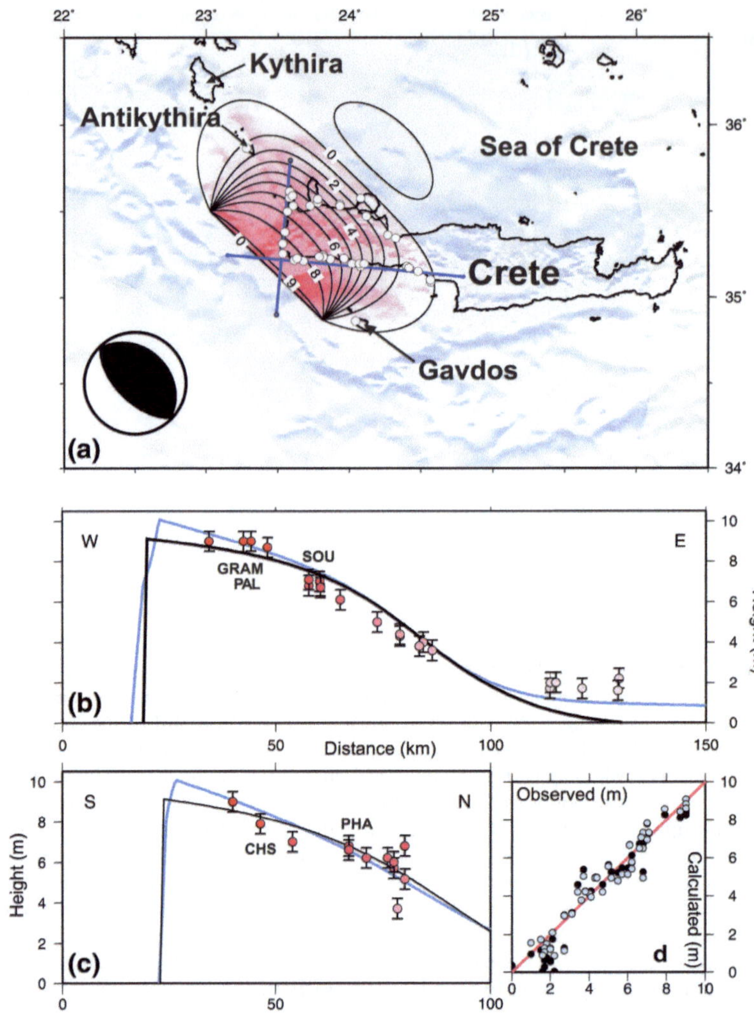

Fig. 2.9 Observed and modelled uplift associated with the AD 365 earthquake. **a** Contours of uplift calculated from the best-fitting elastic dislocation fault model, represented by the *focal mechanism*. *White circles* indicate sites of uplift observations on Crete, Antikythira and Gavdos. **b** W-E profile of shoreline elevations along south coast of Crete measured by Pirazzoli et al. [7], Spratt [4] and ourselves, with 1 m error bars representing uncertainties arising from tidal range and measurement errors. The *black line* shows uplift calculated for best-fitting elastic model for earthquake; the *blue line* shows calculation allowing for post-seismic viscous relaxation (see text). **c** As for (**b**) but for S-N profile along west coast. **d** Observed and modelled elevations of paleo-shorelines for the bestfitting elastic dislocation model, with 20 m of slip (*black circles*) and for 25 m of slip, followed by complete post-seismic relaxation (*blue circles*)

A previously published model of the uplift [21] required a fault dipping at 40° to a depth of 70 km, well below the maximum seismogenic depth of 45 km beneath Crete. I have found solutions that allow slip to such a depth, but regard them as unlikely

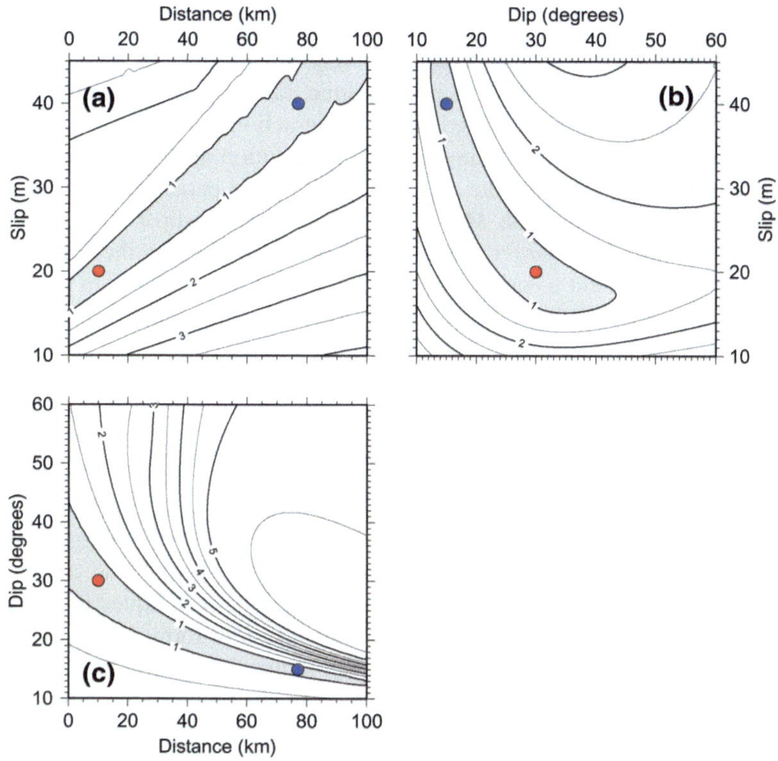

Fig. 2.10 RMS misfit (in metres) between measured and modelled shoreline heights, showing the trade-offs between: **a** slip and the distance of the fault outcrop from the SW corner of Crete; **b** dip of and slip on the fault; **c** dip and the distance of the fault outcrop from the SW corner of Crete. *Red dots* mark the best-fitting solution and *blue dots* show the best-fitting solution for a fault with dip fixed to 15°

to represent what occurred in the AD 365 earthquake, because they are generated only by the need to fit uplift of ~2 m beyond 100 km from the SW corner of Crete (Fig. 2.9b). Post-seismic relaxation after a thrust-faulting earthquake tends to cause uplift above points landward of the base of the co-seismic fault, and I regard it as more likely that the distant uplift is caused by post-seismic relaxation than by deep co-seismic slip. The data do not warrant a sophisticated analysis of post-seismic relaxation, but I have tested the plausibility of this suggestion by a combined co- and post-seismic calculation [22] in which the fault cuts through an upper, purely elastic, layer of 45 km thickness, that lies upon a visco-elastic half-space. This calculation shows that 25 m of slip on the best-fitting fault, followed by complete viscous relaxation (for which case the viscosity of the half-space need not be specified), accounts better for the observations of uplift than does the purely elastic model, particularly in its ability to fit the distant uplift of ~2 m (Fig. 2.9d) (25 m, rather than 20 m, co-seismic slip is required because viscous relaxation causes subsidence close to the fault on the side of the fault that was uplifted in the earthquake). Intriguingly, where uplift is less than 2 m, the algal encrustation that marked the shoreline at higher elevations is replaced by a deep erosive notch (compare Fig. 2.3 with Fig. 2.4). If the uplift here occurred gradually as post-seismic, rather than co-seismic uplift, the encrustation may have been removed by the wave-action.

The parameters of the best-fitting dislocation solution allow me to estimate the minimum seismic moment for the AD 365 event. These parameters are equivalent to an earthquake of magnitude M_w 8.3–8.5. The ratio of slip ($\sim$20 m) to length ($\sim$100 km) for this fault is larger than normally observed [23], but the fault is similar in area and displacement to one of the maximum-slip fault patches in the M_w 9.3 Sumatra earthquake of 2004 [24], and to the 1897 Assam earthquake [25].

2.5 The Tsunami

The tsunami simulations carried out here were performed with ICOM (the Imperial College Ocean Model, http://amcg.ese.ic.ac.uk/ICOM [26]) by Piggott and colleagues. ICOM discretises the Navier-Stokes equations in 3D and assumes the Boussinesq approximation. Finite element discretisation techniques are employed on unstructured meshes, with the discretisation and mesh formed in Cartesian coordinates on the surface of an assumed sphere of radius 6,378 km. The starting point for the tsunami simulation is the construction of an unstructured mesh representing the domain of interest. One minute GEBCO bathymetric data of the entire present-day Mediterranean Sea was optimised using Terreno (http://amcg.ese.ic.ac.uk/Terreno, [27]); the result is a one-element-deep tetrahedral mesh of approximately 2.8×10^5 nodes and 8.2×10^5 elements. No wetting/drying capability was used here, that is, the chaotic run-up that occurs when the tsunami-wave oversteepens and breaks was not modelled, so the mesh stops at the 20 m bathymetric depth contour. A numerical 'tide gauge' was placed at the edge of the mesh approximately 3 km off-shore of Pharos (near Alexandria, Egypt). Crank–Nicolson time-stepping was used with a time-step of 30 s. Continuous piecewise-linear basis functions are used for the spatial discretisation with a second-order anisotropic Smagorinsky LES turbulence model where the filter length scale depends on the local mesh size and shape.

The tsunami calculations were carried out using the best-fitting model of fault slip to provide initial conditions for the submarine surface displacement. Figure 2.11 shows snapshots of the calculated sea surface height at times of 4–90 min after the earthquake, using the surface displacement pattern shown in Fig. 2.9 as the initial condition. The calculation shows that direct waves have the most significant effect, travelling towards the Nile Delta and up the Adriatic. The phases reflected off shorelines interfere and become incoherent in the relatively enclosed Mediterranean Basin.

A prominent feature of the calculation is the wave sweeping east along the African coast to the Nile delta (Fig. 2.11). In AD 365 the offshore island of Pharos (site of the famous lighthouse) was linked to the land by a causeway, the Heptastadion, which was vulnerable to the SW and flooded by the tsunami. It was here that much destruction was concentrated.

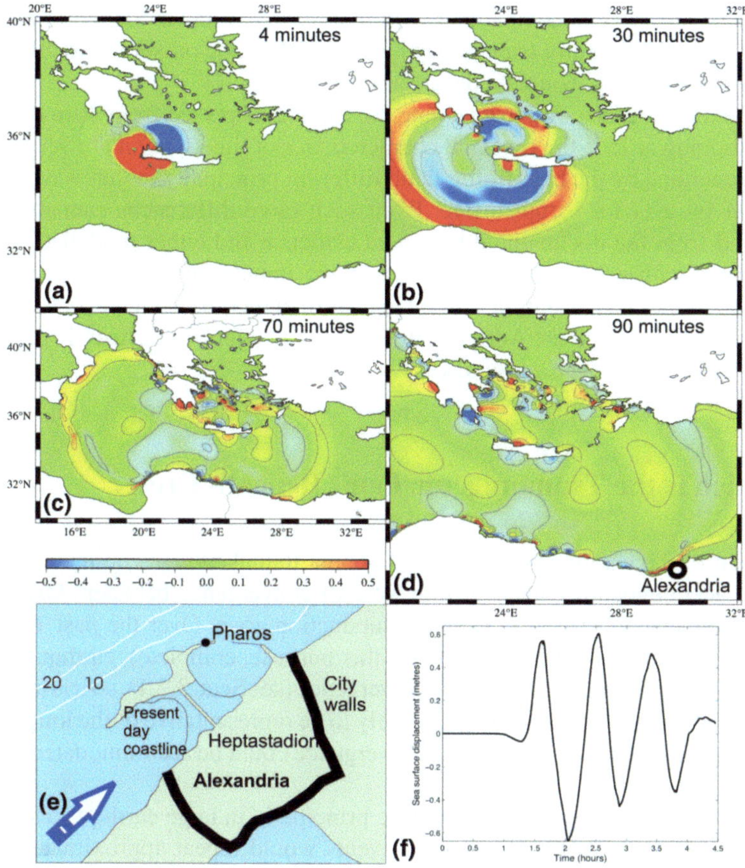

Fig. 2.11 Tsunami calculation. **a–d** Heights of the sea surface at 4, 30, 70 and 90 min after the earthquake. The direct wave travels to the north African coast and into the Adriatic, consistent with the historical record. **e** Sketch map of Alexandria at the time of the AD 365 earthquake. The island of Pharos was connected to the mainland by a narrow causeway, known as the Heptastadion, which was overtopped by the tsunami. Our calculations show that the wave would have arrived from the SW (*white arrow*), explaining how the Heptastadion was inundated despite the apparent protection offered by Pharos Island against a direct wave from Crete (which lies to the NW). **f** Wave height against time that would be observed in the open ocean off Alexandria in water of 20 m depth. The wave height would have been greatly amplified during run-up through shallower water towards the city

Calculation of the sea surface height at a numerical 'tide gauge' situated in the mesh just off the shore of Alexandria shows wave heights of ± 0.6 m in the open ocean (Fig. 2.11f). There are many non-linear effects of the detailed near-shore bathymetry that make it difficult to convert this range into an estimate of the run-up in ancient Alexandria, not least because local conditions of bathymetry and land surface have changed since AD 365 (especially on the Nile Delta). However, the open-ocean amplitude of the wave is comparable with that observed and modelled in the open ocean for the 2004 Sumatra tsunami [28], and therefore the on-shore effects of such a sea wave presumably would be devastating.

A simulation was also run to calculate the tsunami expected from 40 m of slip on the fault dipping at 15°; the essential characteristics of this tsunami are very similar to those generated by the best-fitting fault, differing principally in that wave heights are greater by a factor of up to five. Even such large differences cannot reliably be resolved from the documented historical evidence and either modelled tsunami is compatible with what is known of the AD 365 inundation, with the direct wave reaching all of the coastal sites described in historical records from the Nile delta and north Africa to Sicily and the Adriatic, all of which are much more heavily populated now than at the time of the earthquake.

2.6 Which is the Tsunamigenic Fault Beneath Crete?

Assessment of the significance of the AD 365 earthquake hinges on understanding its role in accommodating the ~35 mm/year of convergence between African and Eurasian lithosphere across the Hellenic subduction zone. Over the past 100 years (in which the record of earthquakes is, for this purpose, complete), earthquakes can account for no more than ~10% of that convergence (see Sect. 3.2.1), so it is necessary to determine whether that percentage is likely to be representative of the longer term, or whether a significant fraction of the convergence could be accommodated by rare, very large, earthquakes.

If the long-term convergence took place principally in large earthquakes like the AD 365 event, then, in SW Crete, such events would repeat approximately every 550 years (for 20 m slip on the steeper fault) or every 1,100 years (for 40 m slip on the plate interface). But field observations and radiocarbon dates show that the 2,000-year-old paleo-shoreline was lifted close to its present position within a few decades of the AD 365 earthquake, implying that this was the *only* earthquake of significant size in that region in the past 1,650 years. Additionally, where the paleo-shoreline is preserved, there is no evidence for previous AD 365-type events since sea-level stabilised near its present height about 6,000 year BP.

This argument can be extended to the whole Hellenic subduction zone, which is about 600 km long. To accommodate convergence across the whole zone by seismic slip in earthquakes would require an earthquake equivalent to the AD 365 event every 100–200 years. The historical record shows a small number of earthquakes in this region (2 or 3 in the past 2,000 years) that could have been of such magnitude but, although the record is incomplete, it is inconceivable that the 10–20 or more great earthquakes that would be needed to remove the seismic slip deficit could have occurred unremarked. For these reasons, I conclude that most of the plate convergence in this region is accommodated aseismically, as did Jackson and McKenzie [29]. This conclusion is reinforced by present-day GPS measurements described later, which show that about 90% of the convergence rate is occurring by steady slip.

If there is some locking along a fault to the south-west of Crete, lines between the Cyclades and Crete should shorten with time. The Cyclades are moving SW at almost the same velocity as Crete, showing that very little locking exists (see below for further analysis).

On the other hand, the occurrence of a magnitude 8.3–8.5 earthquake means that strain must accumulate somewhere. There is no known place where a fault that slips predominantly aseismically also fails in occasional great earthquakes, but a natural way to reconcile the accumulation of elastic strain with aseismic slip on the plate interface would be for that slip rate to be varying with depth. Variation in slip rate would cause strain to accumulate in the surrounding volume, which would eventually be released in an earthquake. One such variation is illustrated in Fig. 2.12, which shows that an increase in slip rate on the plate interface over the depth interval 45–70 km would cause compressional strain to accumulate beneath Crete.

Variation of slip rate with depth might result from a contrast in mechanical properties of different parts of the overriding plate, for instance between the accretionary prism of sediments and the continental crust of the Aegean. The combination of faulting suggested here may be common in subduction zone settings. A thrust fault has been imaged in the Nankai trench off Japan at the landward edge of the accretionary prism [30, 31] and slip on a similar fault is thought to have caused the 10 m uplift of Middleton Island during the 1964 Alaska earthquake [32, 33].

Repeated slip of the kind I propose for the AD 365 earthquake would cause permanent uplift of the surface of Crete by thickening the pile of sediments beneath it [34], and may account for the low seismic wave speeds beneath the island [35]. The contrast between the steep southern coast of Crete and its gentle northern slope (Fig. 2.7, Chap. 5) is consistent with tilting by uplift on a fault cropping out near the south coast, so this proposed fault may control both the uplift and the extent of western Crete itself. Indeed, this geometry of faulting may be a common cause of long-term coastal uplift above subduction zones.

In summary, I conclude that the AD 365 earthquake took place by 20 m of slip on a fault ~100 km long, dipping NE at 30° from the Hellenic Trench to a depth of 45 km. It has long been thought that slip along the plate interface between the Aegean and the Mediterranean ocean floor is predominantly aseismic. My arguments here support that view but emphasise that the Hellenic Trench, which is a major feature along 400 km of the western coast of Greece, appears to mark the outcrop of a separate fault that is capable of producing rare, very large, tsunamigenic earthquakes (see also [8]). In the following section, I attempt to assess the likelihood of a repetition of an event like the AD 365 earthquake.

2.7 Future Tsunamigenic Earthquakes in the Hellenic Subduction Zone

It is possible to estimate the amount of elastic strain accumulating near the Hellenic trench by measuring relative motion across the Southern Aegean using a continuous GPS network. The GPS stations whose data are used here are part of a network of

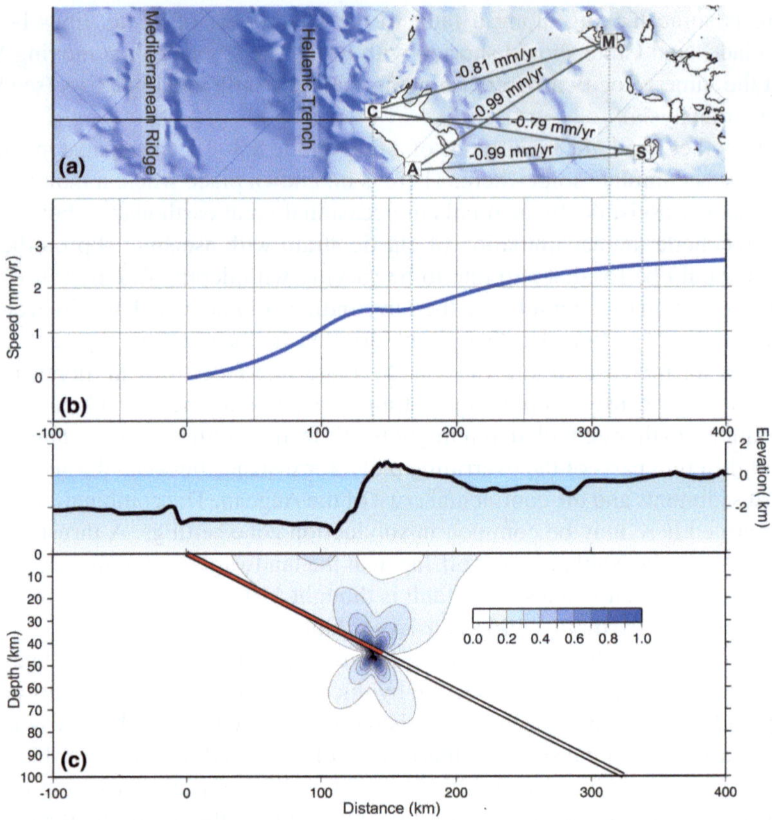

Fig. 2.12 **a** Continuous GPS data show shortening of ~1 mm/year between Chrisoskalitissa (*C*) and Anopoli (*A*) in SW Crete and Milos (*M*) and Santorini (*S*), ~200 km to the NE. **b** Southwestward speed, calculated from model illustrated in (**c**), of points within the Aegean, relative to the Mediterranean Ridge. **c** Distribution of shear stresses calculated [17] for a planar fault, representing the interface between the African plate and the Aegean lithosphere, that slips 4 mm/year more rapidly below 45 km (*open line*) than above that depth (*red line*). Relative values of shear stress are shown by the *colour bar*; absolute values, at any particular time, depend linearly on the slip difference that has accumulated between the two fault segments

20 continuous stations run by the National Technical University of Athens and the University of Oxford. Data was processed by M. Floyd. Lines joining stations in western Crete to Milos and Santorini, about 200 km to the north, are shortening by approximately 0.9 mm/year (Fig. 2.12a), with an uncertainty of about 0.1 mm/year. Such shortening is a small fraction of the total Nubia-Aegean convergence, confirming that the majority of the convergence is taken up by steady sliding along the subduction zone interface. However, this small amount of contraction can be explained by locking on a fault to the south-west of Crete. If we assume that elastic strain builds up because of a variation of slip rate with depth, we can calculate what proportion of the 35 mm/year convergence is going into locking. The observed

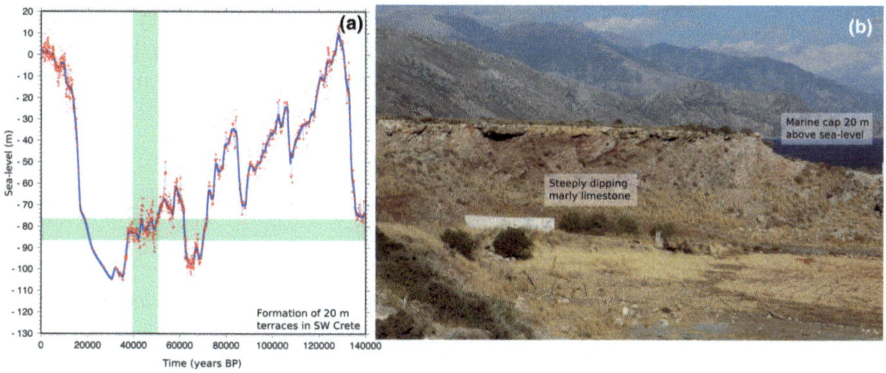

Fig. 2.13 **a** Sea-level curve plotted for the last 140,000 years from Rohling et al. [37]. Small *pink dots* are raw data points, and the *red dots* result from their application of a three-point moving average. The *pale blue line* is my interpolation between these *red points*, using an Akima spline [44]. The *dark blue line* shows the result of applying a Gaussian filter of width 3,000 years to this interpolated curve. The radiocarbon dates of bivalves collected from terraces at 20 m elevation from Paleochora and Chrisoskalitissa suggest that these terraces formed between 40,000 and 50,000 years BP. At this time, sea-level was 75–85 m below present day sea-level (*green band*). **b** Photograph of the 20 m terrace at Paleochora (35° 13.626′N, 23° 40.895′E). The tilted country rock is unconformably overlain by a horizontal marine cap from which the bivalve sample was taken

contraction is consistent with a slip deficit of ~4 mm/yr on the upper part of the subduction interface (Fig. 2.12b); in other words about 10% of the total convergence rate is being stored as elastic strain, with the rest being taken up by aseismic slip. At this rate it would take ~5,000 years to accumulate strain equivalent to the 20 m of slip that was released in the AD 365 earthquake. The predecessor of the AD 365 earthquake would then, plausibly, have happened more than 6,000 years ago, when sea-level was tens of metres below its present position [36, 37].

Any direct evidence of uplift in an earthquake a few thousand years before 6,000 years BP would now be submerged, but some support for this analysis comes from marine terraces at 20–22 m present elevation in western Crete (see Sect. 5.5). Bivalves were collected from the marine encrustation at the top of the terraces, and were radiocarbon dated. These AMS ^{14}C dates show the terraces were formed 40–50,000 years BP (Table 2.1, Fig. 2.13), when the sea surface was ~75–85 m below its present level [36, 37]. Therefore ~100 m uplift has occurred in 40,000–50,000 years. This observation independently suggests a long-term average uplift rate of ~2.0–2.5 mm/yr, which is consistent with uplift occurring in AD 365-type events that repeat every ~4,500 years (4,000–5,000 years). See Chap. 5 for further discussion.

The processes described here for western Crete may also occur along the rest of the Hellenic subduction zone. The AD 1303 earthquake and tsunami are thought to have originated near Rhodes [1, 38], so the whole Hellenic subduction zone may represent a tsunami hazard for the eastern Mediterranean. If the partitioning between aseismic and seismic slip is the same along the whole zone as in the 100 km section near SW Crete, an AD 365-type earthquake should be expected every ~800 years. That there has been only one other such known event (in AD 1303) in the past 1,650 years should focus our attention on the modern-day tsunami hazard in the Eastern Mediterranean.

References

1. N.N. Ambraseys, C.P. Melville, R.D. Adams, *The Seismicity of Egypt, Arabia and the Red Sea: A Historical Review* (Cambridge University Press, Cambridge, 1995)
2. A. Elias, P. Tapponnier, S.C. Singh, G.C.P. King, A. Brias, M. Daeron, H. Carton, A. Sursock, E. Jacques, R. Jomaa, Y. Klinger, Active thrusting off Mount Lebanon: Source of the tsunamigenic AD 551 Beirut-Tripoli earthquake. Geology **35**, 755–758 (2007)
3. A. Salamon, T. Rockwell, S.N. Ward, E. Guidoboni, A. Comastri, Tsunami hazard evaluation of the Eastern Mediterranean: Historical analysis and selected modeling. Bull. Seismolog. Soc. Am. **97**, 705–724 (2007)
4. T.A.B. Spratt, *Travels and Researches in Crete*, vol 2 (J. van Voorst, London, 1865)
5. P.A. Pirazzoli, J. Thommeret, Y. Thommeret, J. Laborel, L.F. Montaggioni, Crustal block movements from Holocene shorelines: Crete and Antikithera (Greece). Tectonophysics **86**, 27–43 (1982)
6. P.A. Pirazzoli, J. Ausseil-Badie, P. Giresse, E. Hadjidaki, M. Arnold, Historical environmental changes at Phalasarna harbor, West Crete. Geoarcheology **7**, 371–392 (1992)
7. P.A. Pirazzoli, J. Laborel, S.C. Stiros, Earthquake clustering in the Eastern Mediterranean during historical times. J. Geophys. Res. **101**, 6083–6097 (1996)
8. B.C. Papazachos, Large seismic faults in the Hellenic arc. Ann. Geophys. **39**, 892–903 (1996)
9. P. England, P. Molnar, Surface uplift, uplift of rocks, and exhumation of rocks. Geology **18**, 1173–1177 (1990)
10. Y. Thommeret, J. Thommeret, J. Laborel, L.F. Montaggioni, P.A. Pirazzoli, Late Holocene shoreline changes and seismo-tectonic displacements in western Crete (Greece). Zeitschrift für Geomorphologie Suppl. **40**, 127–149 (1981)
11. S. Price, T. Higham, L. Nixon, J. Moody, Relative sea-level changes: reassessment of radiocarbon dates from Sphakia and west Crete. Annu. British Sch. Athens **97**, 171–200 (2002)
12. K.A. Hughen, M.G.L. Baillie, E. Bard, J.W. Beck, C.J. Bertrand, P.G. Blackwell, C.E. Buck, G.S. Burr, K.B. Cutler, P.E. Damon, R.L. Edwards, R.G. Fairbanks, M. Friedrich, T.P. Guilderson, B. Kromer, G. McCormac, S. Manning, C. Bronk Ramsey, P.J. Reimer, R.W. Reimer, S. Remmele, J.R. Southon, M. Stuiver, S. Talamo, F. Taylor, J.v.d. Plicht, C.E. Weyhenmeyer, Marine04 marine radiocarbon age calibration, 0–26 cal kyr BP. Radiocarbon **46**, 1059–1086 (2004)
13. P.J. Reimer, F.G. McCormac, Marine radiocarbon reservoir corrections for the Mediterranean and Aegean Seas. Radiocarbon **44**, 159–166 (2002)
14. T. Taymaz, J. Jackson, R. Westaway, Earthquake mechanisms in the Hellenic Trench near Crete. Geophys. J. Int. **102**, 695–731 (1990)
15. J. Mascle, E. Chaumillon, An overview of Mediterranean Ridge collisional accretionary complex as deduced from multichannel seismic data. Geo-Mar. Lett. **18**, 81–89 (1998)
16. B. Tichelaar, L. Ruff, Depth of seismic coupling along subduction zones. J. Geophys. Res. **98**, 2017–2038 (1993)
17. Y. Okada, Internal deformation due to shear and tensile faults in an elastic half space. Bull. Seismolog. Soc. Am. **82**, 1018–1040 (1992)
18. de J.B. Chabalier, H. Lyon-Caen, A. Zollo, A. Deschamps, P. Bernard, D. Hatzfeld, A detailed analysis of microearthquakes in western Crete from digital three-component seismograms. Geophys. J. Int. **110**, 347–360 (1992)

19. T. Meier, M. Rische, B. Endrun, A. Vafidis, H.P. Harjes, Seismicity of the Hellenic subduction zone in the area of western and central Crete observed by temporary local seismic networks. Tectonophysics **383**, 149–169 (2004)
20. P.A. Pirazzoli, The Early Byzantine Tectonic Paroxysm. Zeitschrift für Geomorphologie Suppl. **62**, 31–49 (1986)
21. S. Stiros, A. Drakos, A fault model for the tsunami-associated, magnitude >8.5 Eastern Mediterranean, AD 365 earthquake. Zeitschrift für Geomorphologie **146**, 125–137 (2006)
22. R. Wang, F. Lorenzo-Martin, F. Roth, PSGRN/PSCMP—a new code for calculating co- and post-seismic deformation, geoid and gravity changes based on the viscoelastic-gravitational dislocation theory. Comput.Geosci. **32**, 527–441 (2006)
23. C.H. Scholz, C.A. Aviles, S.G. Wesnousky, Scaling differences between large interplate and intraplate earthquakes. Bull. Seismolog. Soc. Am. **76**, 65–70 (1986)
24. C. Subarya, M. Chlieh, L. Prawirodirdjo, J.-P. Avouac, Y. Bock, K. Sieh, A.J. Meltzner, D.H. Natawidjaja, R. McCaffret, Plate-boundary deformation associated with the great Sumatra–Andaman earthquake. Nature **440**, 46–51 (2006)
25. R. Bilham, P. England, Plateau pop-up in the great (1987) Assam earthquake. Nature **410**, 806–809 (2001)
26. R. Ford, C.C. Pain, M.D. Piggott, A.J.H. Goddard, de C.R.E. Oliveira, A.B. Umpleby, A non-hydrostatic finite-element model for three-dimensional stratified oceanic flows. Part I: model formulation. Monthly Weather Rev. **132**, 2816–2831 (2004)
27. G.J. Gorman, M.D. Piggott, C.C. Pain, de C.R.E. Oliveira, A.P. Umpleby, A.J.H. Goddard, Optical bathymetric representation through constrained unstructured mesh adaptivity. Ocean Model. **12**, 436–452 (2006)
28. J. Gower, Jason 1 detects the 26 December 2004 tsunami, EOS transactions. AGU **86**, 37–38 (2005)
29. J. Jackson, D. McKenzie, The relationship between plate motions and seismic moment tensors, and rates of active deformation in the Mediterranean and Middle East. Geophys. J. **93**, 45–73 (1988)
30. J.-O. Park, T. Tsuru, S. Kodaira, P.R. Cummins, Y. Kaneda, Splay fault branching along the Nankai subduction zone. Science **297**, 1157–1160 (2002)
31. G.F. Moore, N.L. Bangs, A. Taira, S. Kuramoto, E. Pangborn, H.J. Tobin, Three-dimensional splay fault geometry and implications for tsunami generation.. Science **318**, 1128–1131 (2007)
32. G. Plafker, Tectonic deformation associated with the 1964 Alaska earthquake. Science **148**, 1675–1687 (1965)
33. G. Ichinose, P. Somerville, H.K. Thio, R. Graves, D. O'Connell, Rupture process of the 1964 Prince William Sound, Alaska, earthquake from the combined inversion of seismic, tsunami and geodetic data. J. Geophys. Res. **112**, B07306 (2007)
34. X. Le Pichon, J. Angelier, The Aegean Sea. Philos. trans. R. Soc. Lond.. Ser. A. Math. Sci. **300**, 357–372 (1981)
35. B. Endrun, T. Meier, M. Bischoff, H.-P. Harjes, Lithospheric structure in the area of Crete constrained by receiver functions and dispersion analysis of Rayleigh phase velocities. Geophys. J. Int. **158**, 592–608 (2004)
36. M. Siddall, E. Rohling, A. Almogi-Labin, C. Hemleben, D. Meischner, I. Schmelzer, D. Smeed, Sea-level fluctuations during the last glacial cycle. Nature **423**, 853–858 (2003)
37. E.J. Rohling, K. Grant, M. Bolshaw, A.P. Roberts, M. Siddall, C. Hemleben, M. Kucera, Antarctic temperature and global sea-level closely coupled over the past five glacial cycles. Nat. Geosci. **2**, 500–504 (2009)
38. A.Z. Hamouda, Numerical computations of 1303 tsunamigenic propagation towards Alexandria, Egyptian Coast. J. Afr. Earth Sci. **44**, 37–44 (2006)
39. H. Akima, A method of bivariate interpolation and smooth surface fitting for irregularly distributed data points. ACM Trans. Math. Softw. **4**, 148–159 (1978)
40. S. McClusky, Global positioning system constraints on plate kinematics and dynamics in the eastern Mediterranean and Caucasus. J. Geophys. Res. **105**, 5695–5719 (2000)

41. S. McClusky, R. Reilinger, S. Mahmoud, D. Ben Sari, A. Tealeb, GPS constraints on Africa (Nubia) and Arabia plate motions. Geophys. J. Int. **155**, 126–138 (2003)
42. R. Reilinger, S. McClusky, P. Vernant, S. Lawrence, S. Ergintav, R. Cakmak, H. Ozener, F. Kadirov, I. Guliev, R. Stepanyan, M. Nadariya, G. Hahubia, S. Mahmoud, K. Sakr, A. Ar-Rajehi, D. Paradissis, A. Al-Aydrus, M. Prilepin, T. Guseva, E. Evren, A. Dmitrotsa, S.V. Filikov, F. Gomez, R. Al-Ghazzi, G. Karam, GPS constraints on continental deformation in the Africa–Arabia–Eurasia continental collisional zone and implications for the dynamics of plate interactions. J. Geophys. Res. **111**, B05411 (2006)
43. E.R. Engdahl, R.D. Vander Hilst, R.P. Buland, Global teleseismic earthquake relocation with improved travel times and procedures for depth determination. Bull. Seismolog. Soc. Am. **88**, 722–743 (1998)
44. M.L. Jost, O. Knabenbauer, J. Cheng, H.-P. Harjes, Fault plane solutions of microearthquakes and small events in the Hellenic arc. Tectonophysics **356**, 87–114 (2002)
45. K.A. Hughen, J. Southon, S. Lehman, C. Bertrand, J. Turnbull, ^{14}C calibration and activity for the past 50,000 years updated from the Cariaco Basin. Q. Sci. Rev. **25**, 3216–3227 (2006)

Chapter 3
Earthquakes in the Eastern Mediterranean

3.1 Summary

In this chapter, improved focal mechanisms and centroid depth estimates of earthquakes, combined with GPS velocities, are used to examine the tectonics of the Hellenic subduction zone and in particular the processes occurring at both ends of the Hellenic Arc. Nubia-Aegean convergence is accommodated by shallowly dipping thrust-faulting along the subduction-zone interface, as well as by steeper splay faults in the overriding material. From a comparison of observed and expected seismic moment release over the last 100 years, combined with existing knowledge of the longer-term documented historical record, I confirm earlier suggestions that most (80%) of this convergence is accommodated aseismically, i.e. that the subduction zone is uncoupled. This conclusion is robust, even allowing for rare very large earthquakes on splay faults, such as that of AD 365, and also allowing for the contribution of small earthquakes. The down-going Nubian plate deforms by arc-parallel contraction at all depths, from 200 km seaward of Crete to at least 100 km within the subducting slab. Extensional (T) axes of earthquakes are aligned down-dip within the descending slab suggesting that, even if the aseismic prolongation of the slab has reached the 670 km mantle discontinuity, it is not transmitting stresses to shallower depths. Shallow thrust-faulting earthquakes on the subduction interface show a divergence of slip vectors round the arc, and GPS measurements show that this is accommodated mainly by E-W extension on normal faults in the overriding Aegean material. The eastern end of the subduction zone, south of Rhodes, displays distributed deformation in the overriding material, including a mixture of strike-slip and splay-thrust faulting, and probably involves rotations about a vertical axes. Here slip on the interface itself is by thrust faulting with slip vectors oblique to the arc but parallel to the overall Nubia-Aegean convergence: there is no evidence for slip-partitioning in the traditional sense. In the west, the subduction zone terminates in a distributed zone of parallel NE-SW strike-slip faults, of which the most prominent is the Kefalonia Transform Fault (KTF). A flexural gravity anomaly confirms that the deep bathymetric escarpment of the KTF is a lateral ramp, formed as the Ionian

B. Shaw, *Active Tectonics of the Hellenic Subduction Zone*, Springer Theses,
DOI: 10.1007/978-3-642-20804-1_3, © Springer-Verlag Berlin Heidelberg 2012

islands are emplaced SW onto the Apulian lithosphere, and enhanced by minor thrust faulting with slip vectors perpendicular to the scarp. Distributed parallel strike-slip faults both SW and NE of mainland central Greece terminate in E-W graben in central Greece, which accommodate the overall NE-SW shear by clockwise block rotation. Central Greece therefore acts as a relay zone between the strike-slip faulting of the NE Aegean and the Ionian Islands–Western Peloponnese.

3.2 Introduction

Mediterranean sea floor, of probable Mesozoic age [1, 2], subducts northwards beneath Crete at a rate (35 mm/year) that greatly exceeds the convergence between Africa (Nubia) and Eurasia (5–10 mm/year) because of the rapid SW motion of the southern Aegean itself, relative to Eurasia [3, 4]. The surface morphology of the subduction system is obscured by a sedimentary section up to 10 km thick overlying the ocean crust, which is deformed in a broad accretionary prism south of Crete known as the Mediterranean Ridge (Fig. 3.1, [5–9]). A dramatic 2–3 km high, south-facing bathymetric scarp runs in an arc between the Peloponnese and Crete, splitting into at least three branches south of Crete, and continuing east to Rhodes as the Ptolemy, Pliny and Strabo trenches (Fig. 3.1, [5, 10, 11]). Although this scarp is referred to here, and by others, as the Hellenic Trench, it is clearly not the expression of an oceanic trench in the usual sense, as the earthquakes directly beneath it are low-angle thrusts at depths of 20–40 km [12], and it is not the site where those thrusts would project to the surface. Assuming those thrusts represent the interface between the down-going African lithosphere and the overriding Aegean, that interface would instead project to the surface at least 150 km south of Crete, beneath the aseismic Mediterranean Ridge. In fact, it is more likely that the interface never reaches the surface at the Mediterranean Ridge at all, but is 'blind', and dissipates the motion in sediment thickening and folding instead. The escarpment of the Hellenic Trench may thus be a backstop to the deformed accretionary prism, where sea floor sediments that are decoupled aseismically from the oceanic crust by weak layers, including Messinian salt and lower Cretaceous shales [13, 14], abut against the continental crust of Crete and the Aegean (e.g. [6, 15]). It is probable that the scarp of the Hellenic Trench is the surface expression of a steep ($\sim$30°) reverse fault splaying off the deeper underlying thrust-fault interface of the subduction zone (see Chap. 2).

Where surface features are obscured or complicated, the obvious recourse is to look at earthquake focal mechanisms. The Global CMT catalogue and other compilations (e.g. [16–18]) show a bewildering variety of mechanisms associated with the subduction: normal, thrust and strike-slip fault-plane solutions with a range of orientations all apparently in nearly the same place. Taymaz et al. [12] showed that the way to clarify this information was through careful quality control of the earthquake data. Using body-waveform modelling to improve and constrain focal mechanisms and, especially, centroid depths, they exposed a clear pattern, revealing that the earthquakes near Crete split into three groups. The Nubia-Aegean

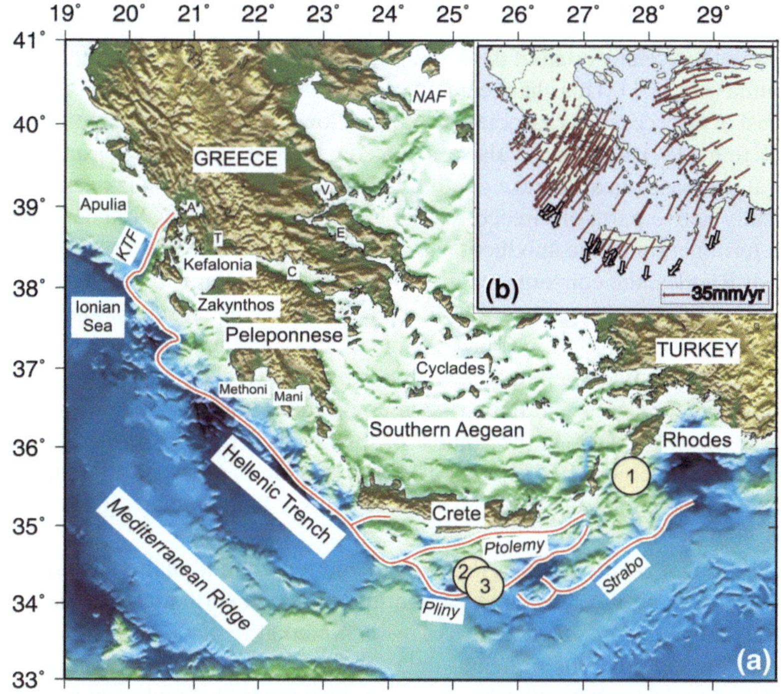

Fig. 3.1 **a** Principal tectonic features and bathymetry–topography [99, 100] of the Hellenic Subduction Zone. The Hellenic Trench is a 4 km high escarpment, which branches into three deep trenches in the east, and terminates in the KTF in the west. The Mediterranean Ridge is a thick (up to 10 km) pile of sediment forming the accretionary prism of the subduction zone. The locations of the three example earthquakes in Sect. 3.3.4 and Sect. 3.3.5 are represented by *yellow circles*. The extensional graben of Central Greece are labelled as letters: *C* Corinth, *E* Evia, *I* Arta, *V* Volos, *T* Trichonis Lake. **b** *Red arrows* are GPS velocity vectors in the region relative to Nubia (see text). Slip vectors of thrust-faulting earthquakes along the subduction interface and on splay thrust faults in the overriding material are plotted as *white arrows*

convergence itself is achieved by low-angle (10–20°) thrust faults with steeper reverse-fault splays at shallower depths above them. The overriding Aegean crust has mechanisms that involve arc-parallel extension, mostly by normal faulting. The underthrusting African lithosphere undergoes arc-parallel shortening, achieved by both thrust and strike-slip faulting.

A lack of earthquakes elsewhere restricted the Taymaz et al. [12] study to the area around Crete. Since then, the greatly increased earthquake data set, combined with excellent GPS data, potentially allow us to see whether the pattern found by Taymaz et al. [12] near Crete is representative of the whole arc, and to interpret the earthquake focal mechanisms, depths, slip vectors and P and T axes in their geodynamical context. These are the broad aims of this chapter. One interest concerns the extent to which the pattern itself, in general, is related to the strong curvature of the arc. Particular questions are associated with the western and eastern ends. In the west, the subduction zone terminates where the down-going side changes along strike from the oceanic crust of the Ionian Sea to the continental crust of the Adriatic and Apulia (Fig. 3.1). This change involves a right-step from the Ionian islands to the NW coast of Greece, and is associated with a major strike-slip fault, sometimes called the KTF (e.g. [19]), though it is also associated with thrust faulting, coastal uplift and a general decrease of earthquake depths. This western termination is one of the most seismically active parts of Greece, yet its kinematics are not well understood. The eastern termination of the subduction zone is also obscure. GPS observations show that, near Rhodes, the convergence direction is highly oblique to the Ptolemy, Pliny and Strabo trenches, suggesting that the zone must involve a substantial strike-slip component to the convergence (Fig. 3.1), but a lack of earthquakes has left this region almost untouched by earlier studies, beyond speculation about possible partitioning of the convergence onto parallel strike-slip and thrust faults, as is indeed common elsewhere in the oceans. (In the context of this thesis, I take 'partitioning' to mean that oblique convergence is separated into perpendicular strike-slip and thrusting components, which are taken up on parallel faults.) These kinematic questions are primary foci of this chapter.

An additional oddity of the Hellenic subduction zone is that it is known to be largely aseismic, in the sense that there are not enough earthquakes to account for the Africa-Aegean convergence [20]. This is hinted at by an assessment of the last 100 years' quantitative instrumental data, which accounts for a maximum of 10–15% of the known convergence, but becomes even more likely when combined with the known historical record. The 100 years' seismic deficit requires an earthquake of M_w 8.0 every 14 years, or one of M_w 7.1 every year, to rectify the situation; there is simply no suggestion of this extreme level of activity in the historically documented record [21–23]. Yet the historical record *does* contain two earthquakes of $M_w > 8$ in the last 2,000 years, each of which must have had a source dimension of greater than 100 km [24–28]. These two earthquakes are not nearly large enough or frequent enough to account for the missing seismic moment: together they can only account for an extra 5–10% of the expected moment over 2,000 years. But they raise a paradox: it is easy to understand how a largely aseismically slipping interface may produce occasional earthquakes of M_w 6, with source dimensions of about 10 km, on a few locked patches; but it is harder to see how the same interface can be both 75% aseismic and also be capable of accumulating strain over a sufficiently large area to generate earthquakes of $M_w > 8$ [29, 30].

In Chap. 2, I suggested a resolution of this paradox, in which rare M_w 8 earthquakes are generated on high-angle splays off the main subduction interface at depth. These splay faults account for 5–10% of the convergence, while the bulk (90%) of the convergence is accommodated up-dip on the main interface itself, nearly all of it by aseismic slip with small local patches slipping in earthquakes of up to about M_w 6.0. In this description only the 5–10% of the convergence accommodated by the high-angle splays is fully coupled and stores significant elastic strain; the remainder is largely uncoupled. This conclusion is supported by GPS measurements that show only 1 mm/year of shortening between Crete and the Cyclades. (These arguments are quantified further in Sect. 3.2.1 where I show that Ganas and

Parson's [31] claim that the Hellenic subduction zone is fully coupled cannot be correct.) In Chap. 2, I suggested also that such splays account for the presence of the Hellenic Trench bathymetric escarpment itself, as well as the uplift and tilting of Crete. By contrast, slip along the subduction zone interface can only produce the observed AD 365 uplift if slip extends to implausibly great depths (around 70 km), and if the fault slips by improbably large amounts (40–50 m), as shown by Stiros and Drakos [27], Ganas and Parsons [31] and in Chap. 2. If correct, my interpretation of the AD 365 event and the faulting near Crete has important implications for the earthquake and tsunami hazard of the eastern Mediterranean, and other large-earthquake-prone regions of the subduction zone could be identified by examination of the earthquake focal mechanisms and depths, and of the geomorphology of any adjacent land.

The bulk of this chapter is concerned with detailed examination of earthquake focal mechanisms, depths and slip vectors and their comparison with GPS observations. These are considered round the arc in groups (a) within the subducting lithosphere, (b) on the interface, and (c) within the overriding Aegean. Special consideration is given to the western and eastern subduction zone terminations, followed by a summary of the implications of the study for the overall tectonics of the region.

3.2.1 Earthquake Deficit Calculation

It is possible to calculate the expected seismic moment release rate in the Hellenic subduction zone if it is fully coupled using the known rate of convergence between Nubia and the Southern Aegean. The total along-strike length of the arc is 600 km, and the convergence rate is 35 mm/year [4]. Focal mechanisms of earthquakes in Fig. 3.7 show that the subduction zone interface dips at 15° and that the deepest earthquakes on it occur at a depth of 40 km. The down-dip width of the seismogenic interface is therefore 40 km/sin 15°, or 155 km. Since seismic moment $M_o = \mu A s$, where μ is the rigidity modulus (3×10^{10} Pa s), A is the area of the fault plane, and s is slip along the fault plane, the moment release rate that would be expected if earthquakes accommodated all of the Nubia-Aegean motion is $\sim 97 \times 10^{18}$ Nm/year. Several authors [20, 31] have estimated the rate of moment released in earthquakes, averaged over the last 100 years or fewer, to be between 3×10^{18} Nm/year and 19×10^{18} Nm/year, which is much less than the above fully-coupled requirement.

Here, these estimates of the seismically released moment over the last 100 years are checked, updated and refined. The catalogue of Jackson and McKenzie [20] is claimed to be complete above M_w 6.0 between 1908 and 1981, but the locations and depths of the earlier earthquakes in this catalogue are not well-determined. Jackson and McKenzie [20] separated earthquakes that occurred near the Hellenic trench into two categories: those with probable shallow focal depths (which might contribute to the convergence), and earthquakes with deeper focal depths that are probably related to deformation within the subducting slab. Even if only the apparently shallow earthquakes (Table 3.1) are considered, the moment release obtained is likely to be an upper bound on the true moment release, because it is now known that some of them accommodate along-strike shortening or extension of the down-going or overriding material and not convergence in the subduction zone itself. If one extends the catalogue to 2009, and then allows for the additional contribution of earthquakes with $M_w < 6.0$ [32], an estimate of the observed average seismic moment release rate over the last 100 years is between 7.4×10^{18} Nm/year and 15.8×10^{18} Nm/year, depending on whether only the apparently shallow earthquakes (Table 3.1), or all of those from the catalogue of Jackson and McKenzie [20], including those that might be deep (Table 3.2), are included in the calculation. This equates to between 8 and 16% of the Nubia-Aegean motion, confirming the substantial seismic moment deficit inferred by earlier authors.

Thus, over the last 100 years, earthquakes have been unable to account for at least 85% of the moment required to accommodate the known Nubia-Aegean convergence. If this deficit is to be made up by occasional big earthquakes it would require an earthquake of M_w 7.1 every year, or an earthquake of M_w 8.0 earthquake every 14 years. Only 2 events of $M_w \sim 8$ are known in the last 2,000 years, in AD 365 and AD 1303. Even if an AD 365-type event of M_w 8.3–8.5 occurs every 800 years, which may be possible based on the analysis in Chap. 2, the moment release rate from such large events is only $\sim 5.5 \times 10^{18}$ Nm/year. These rare very large events may therefore take up another 5–10% of the moment release, but it is inconceivable that the remaining 75% or more could have been released seismically without being noticed in the region's rich historical record [22, 23].

GPS velocities in the region also confirm that the subduction zone interface must be largely uncoupled: the Cyclades are moving towards Nubia at only 1 mm/year faster than western Crete (see Chap. 2, Fig. 2.12). If the subduction zone interface were locked, there would be a much greater difference in velocity, since Crete would be attached to the down-going plate and there would be contraction of $\sim$30 mm/year between Crete and the Cyclades.

3.3 Data and Methods

3.3.1 Earthquake Epicentres

To make correct associations of earthquakes with tectonic features it is important to use epicentres that are as accurate as possible. Except for the most recent events, for which only USGS PDE epicentres are available, I use epicentres determined in Engdahl et al. [33] and their updated catalogues (ISC locations are used for events before 1964). With the obviously complicated velocity structure of a subduction zone, and the uneven station distribution contrasting dense coverage in Europe and Asia to the north with sparse coverage in Africa and the oceans to the south, epicentre bias and inaccuracy are potential issues [34, 35].

Table 3.1 Seismic moment released in the Hellenic subduction zone by shallow earthquakes in the last 100 years. 1: Jackson and McKenzie [20], 2: McKenzie [3], 3: This study

Date	Latitude	Longitude	Assumed $M_o \times 10^{19}$ Nm	Location	Source
1912.01.24	38.20	20.50	2.7	Crete shallow	1
1914.11.27	38.50	20.50	0.4	Crete shallow	1
1915.01.27	38.50	20.50	0.7	Crete shallow	1
1915.08.07	38.50	20.50	1.4	Crete shallow	1
1919.02.24	36.70	21.00	0.7	Crete shallow	1
1922.08.11	34.90	27.20	1.0	Crete shallow	1
1922.08.13	35.50	27.80	1.9	Crete shallow	1
1926.03.18	35.90	29.90	3.9	Crete shallow	1
1926.06.26	36.00	28.00	5.5	Crete shallow	1
1927.07.01	36.70	22.40	0.5	Crete shallow	1
1937.12.16	35.00	23.50	0.5	Crete shallow	1
1940.02.29	35.50	25.50	0.2	Crete shallow	1
1947.08.30	35.10	23.40	0.7	Crete shallow	1
1947.10.06	36.80	21.80	2.7	Crete shallow	1
1948.02.09	35.50	27.20	7.7	Crete shallow	1
1948.04.22	38.70	20.40	1.4	Crete shallow	1
1948.06.30	38.90	20.50	0.7	Crete shallow	1
1948.07.24	34.60	24.50	0.5	Crete shallow	1
1952.12.17	24.50	24.40	1.0	Crete shallow	1
1953.08.09	38.10	20.70	0.4	Crete shallow	1
1953.08.11	38.10	20.70	1.0	Crete shallow	1
1953.08.12a	38.10	20.70	7.7	Crete shallow	1
1953.08.12b	38.10	20.70	0.3	Crete shallow	1
1953.10.21	28.30	20.80	0.3	Crete shallow	1
1955.09.12	32.20	29.60	0.3	Crete shallow	1
1957.04.24	36.10	29.00	1.0	Crete shallow	1
1957.04.25	36.50	28.80	5.5	Crete shallow	1
1958.08.27	37.45	20.72	0.7	Crete shallow	1
1959.05.14	35.00	24.80	0.2	Crete shallow	1
1959.11.15	37.83	20.47	5.5	Crete shallow	1
1965.04.05	37.70	21.80	0.2	Crete shallow	1
1969.01.14	36.18	29.20	0.4	Crete shallow	1
1969.06.12	34.40	25.06	0.2	Crete shallow	1
1972.09.17	38.28	20.34	1.0	Crete shallow	1
1976.05.11	37.56	20.35	0.7	Crete shallow	1
1984.06.21	35.38	23.28	0.1	Interface	2
2008.02.14a	36.24	21.79	1.4	Interface	3
2008.02.14b	36.29	21.67	0.2	Interface	3
TOTAL			61.2	Shallow	

To assess the scale of epicentral mislocation, I found six earthquakes that were located both by the global teleseismic network and by dense temporary local

Table 3.2 Other earthquakes in the Hellenic subduction zone considered in the moment deficit calculation from the catalogue of Jackson and McKenzie [20]

Date	Latitude	Longitude	Assumed $M_o \times 10^{19}$ Nm	Location	Source
1918.07.16	35.5	25.5	5.5	Crete deep	1
1923.08.01	35.0	25.0	3.9	Crete deep	1
1923.08.03	35.0	25.0	0.3	Crete deep	1
1925.07.06	38.2	21.7	3.9	Crete deep	1
1926.07.05	36.5	27.0	0.4	Crete deep	1
1926.08.30	36.7	23.2	21.6	Crete deep	1
1927.07.01	36.7	22.7	15.3	Crete deep	1
1929.03.27	36.7	26.5	0.4	Crete deep	1
1930.02.14	35.7	24.7	2.7	Crete deep	1
1934.11.29	36.7	25.7	0.5	Crete deep	1
1935.02.25	35.7	25.0	7.7	Crete deep	1
1935.03.18	35.5	27.0	1.0	Crete deep	1
1937.12.16	35.0	23.5	0.5	Crete deep	1
1941.12.13	37.2	28.3	1.0	Crete deep	1
1942.05.21	37.5	20.5	0.3	Crete deep	1
1942.06.21	36.5	27.0	0.7	Crete deep	1
1943.01.07	37.3	20.6	0.3	Crete deep	1
1943.06.27	35.0	26.0	0.3	Crete deep	1
1943.10.16	36.4	27.9	1.4	Crete deep	1
1944.05.27	36.0	27.5	0.4	Crete deep	1
1945.09.02	33.7	28.5	0.7	Crete deep	1
1947.07.07	38.0	20.5	0.4	Crete deep	1
1953.02.14	35.5	26.5	0.4	Crete deep	1
1953.06.23	36.0	25.0	0.3	Crete deep	1
TOTAL			131.1	Shallow and deep	

seismograph networks in central Greece, the Peloponnese and Crete [36–38]. A comparison between the two locations for each earthquake shows a difference of less than 12 km in all cases, with no consistent pattern in the azimuths of offset (Table 3.3). A distance of 10–15 km is roughly equivalent to the rupture length of a M_w 6.0 earthquake. The size of symbols used to represent earthquakes in maps in this chapter is large compared with this uncertainty in epicentral location, so the true location is likely to lie within the symbols. My subsequent interpretations are not sensitive to errors of this scale.

3.3.2 Focal Mechanisms, Slip Vectors and Depths

The primary interest of this chapter is in the focal mechanisms and depths of earthquakes larger than M_w ~5.2. The best constraints on these parameters come

Table 3.3 Offset between EHB locations and those obtained from dense local networks of stations

Date	Time	Latitude	Longitude	Mw	Δd(km)	Offset azimuth °from N	Source
971105	12:22:56	34.8	24.0	4.5	1	117°	Meier et al. [38]
880712	02:26:53	38.7	23.4	4.4	6	123°	Hatzfeld et al. [37]
880727	05:00:13	35.4	24.8	4.1	11	186°	Hatzfeld et al. [37]
860608	04:54:52	36.0	21.5	5.3	10	131°	Hatzfeld et al. [36]
860614	17:50:22	36.0	22.0	4.7	11	91°	Hatzfeld et al. [36]
860619	15:14:45	36.6	21.2	4.7	11	30°	Hatzfeld et al. [36]
860705	09:52:45	37.9	22.5	4.2	7	277°	Hatzfeld et al. [36]

from long-period teleseismic body-wave modelling, which can provide considerable improvements over routine global CMT catalogue [39] determinations, especially for centroid depths [40–42].

In my modelling procedure I used P and SH body waves recorded by stations of the Global Digital Seismic Network (GDSN). Seismograms in the teleseismic distance range of 30–90° were first convolved with a filter that reproduces the bandwidth of the old WWSSN 15–100 long-period instruments. At these periods the source of earthquakes of $\sim M_w$ 6 appears as a point in space (the centroid) with a finite rupture time, and the seismograms are sensitive to the source parameters of the centroid while relatively insensitive to the details of geological structure. I then used the MT5 implementation [43] of the algorithm of McCaffrey and Abers [44] and McCaffrey et al. [45], which inverts the P and SH waveform data to obtain the strike, dip, rake, centroid depth, seismic moment and the source time function, which is parameterised by a series of isosceles triangle elements of half-duration τ s. Wherever possible I used the observed onset time on the high-frequency broad-band records to align the observed and synthetic waveforms. I always constrained the source to be a double-couple. Stations are weighted by azimuth density, and then the weights of SH waveforms are halved to compensate for their generally larger amplitudes. The method and approach I used are described in detail elsewhere [46–49] and are too routine to justify detailed repetition here.

A simple source-region velocity structure was specified, using up to a maximum of two layers and a water layer. This permits a different average velocity above the source (which determines the delay between direct and surface-reflected arrivals, and hence depth) from that below the source (which controls station positions on the focal sphere). The source itself is always in the underlying half-space. To comply with these requirements, the source structures I chose followed the approach of Taymaz et al. [12] to ensure direct comparison with their results. For the events that are relatively shallow, typically around 20 km, and south of the Hellenic Trench escarpment, I used a structure consisting of a layer 8 km thick with V_p 4.5 km/s, representing the sediment covering the sea floor, overlying a half-space of V_p 6.5 km/s. For those with a centroid depth around 40 km or greater, I used a 24 km thick layer of V_p 6.5 km/s overlying a half-space of V_p 7.8 km/s. For all offshore earthquakes I included a water layer of depth appropriate to their [33] locations. As Taymaz et al.

[12] showed, small variations in the velocity structure used do not significantly affect the focal mechanism and are responsible for depth uncertainties of up to ±4 km. I am therefore confident in assimilating mechanisms and depths from other similar waveform-based studies where slightly different velocity structures have been used (e.g. [18]).

The earthquake mechanisms and depths I discuss in this chapter are ranked in quality by category. The best (in red) are those whose source parameters are independently determined by long-period body-wave modelling and with good P and SH waveform coverage. The formal errors associated with the minimum-misfit solution obtained by the inversion algorithm underestimate the true uncertainties in the source parameters (e.g. [48]). A better estimate of uncertainties is found by fixing some of the source parameters at values close to but different from those of the minimum misfit solution, and seeing whether the match of the observed to synthetic seismograms deteriorates [40, 48, 49]. Such sensitivity analyses showed that for the best-determined earthquakes (in red) the uncertainties are typically ±15° in strike, ±15° in dip, ±20° in rake, and ±4 km in depth. Earthquakes with less waveform data, or with inferior station coverage, but whose centroid depths are nonetheless well-constrained by body-wave modelling are shown in pink, often adopting the best double-couple CMT focal mechanism. In grey I include best double-couple CMT mechanisms of earthquakes for which there is insufficient data, or for which the data is of too poor quality, to carry out independent waveform modelling at all; for these earthquakes I have no accurate depth control. Finally, some of the most reliable first-motion fault-plane solutions, determined from long-period records of the WWSSN in earlier studies by McKenzie [3, 10] are also plotted to extend my coverage back in time to the early 1960s; but for these earthquakes also there is no accurate depth control. The focal mechanism data and, where available, the improved centroid depth estimates, for all earthquakes plotted in this chapter are listed in tables in Chap. 7. Full solutions of waveform-modelled earthquakes are shown in Chap. 8.

I illustrate below (Sects. 3.3.4, 3.3.5) the importance of this careful forensic approach to evaluating focal mechanisms and depths with three examples. In each case the accurate mechanisms and depths reveal or confirm the tectonic significance of the earthquakes, which could otherwise easily be misinterpreted, or would not be known with confidence.

3.3.3 GPS Data

The GPS velocities used here are taken from the data sets of Reilinger et al. [4] (Greece only) and Hollenstein et al. [50], which have been combined with the velocities obtained by the COMET group (personal communication, M. Floyd). For most of the purposes of this chapter, these velocities have been rotated into a Nubian frame of reference (after [51]).

3.3.4 Example 1: Eastern Hellenic Arc, 15 July 2008

On 15 July 2008, a strike-slip earthquake of M_w 6.2 occurred in the eastern part of the subduction zone, south of Rhodes (see Fig. 3.1). In this region, the trend of bathymetric scarps offshore is oblique rather than perpendicular to the direction of convergence between the Aegean and Africa (Fig. 3.1), suggesting two possible fault configurations that could achieve the overall motion: either partitioning of the oblique left-lateral convergence onto separate pure thrust and strike-slip faults, or oblique slip on the main subduction zone interface.

The synthetic and observed waveforms for this event are shown in Fig. 3.2. The nodal planes are well-determined by the available P and SH station distribution, and the depth is well-constrained by the clear surface reflections in both P and SH waves. The left-lateral nodal plane strikes E-W, rather than NE-SW and parallel to the strike of the subduction zone at this point, as would be expected for a partitioned strike-slip component of oblique convergence. In addition, the depth of 56 ± 4 km suggests the event is in the downgoing African lithosphere, as I show later that the subduction interface itself is only seismogenic to about 40–45 km depth.

The P axis for this earthquake does trend NE-SW, and in fact this event is one of several within the downgoing African lithosphere that has a P axis parallel to the strike of the subduction zone. This is a pattern noticed near Crete by Taymaz et al. [12] that, I show later, continues round the whole arc, and may be related to the strong curvature imposed on the downgoing African plate. The reason I have confidence in attributing this earthquake to that group is because I have confidence in the focal mechanism and depth.

3.3.5 Examples 2 and 3: Shallow and Steep Thrust Faulting Earthquakes

An important feature of the Hellenic subduction zone is that the clear bathymetric feature known as the Hellenic Trench is not the surface expression of the main subduction interface. The true geometry and location of that interface is obscured beneath thick piles of sediment which deform and slide down bathymetric gradients. I show in Sect. 3.5 that the larger earthquakes on the subduction interface are at depths of 15–45 km beneath the Hellenic Trench system, similar to the 20–40 km deep interface inferred to exist from microearthquakes [38] and larger earthquakes [12] beneath western and central Crete. Seismograms from one such earthquake (Example 2), which occurred on 28 March 2008 to the south of Crete (Fig. 3.1), are plotted in Fig. 3.3 (first two rows). The figure offers a visual comparison of the quality of the fit to the waveforms for synthetics generated at the best-fitting depth (43 km) and at a much shallower depth (9 km) for the best-fitting fault-plane solution. The fit to the waveforms is severely degraded if the centroid depth is forced to be shallow.

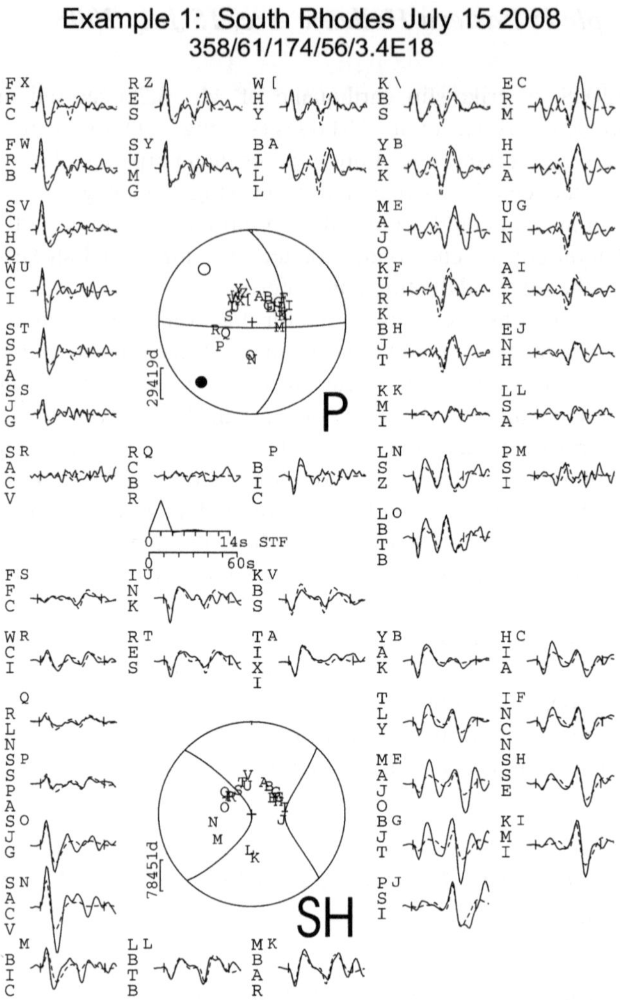

Fig. 3.2 P and SH waveforms for Example 1: south of Rhodes, 15 July 2008. Although it has a strike-slip mechanism, it is not related to the oblique convergence between Nubia and the Aegean, but rather represents along-arc shortening in the down-going Nubian plate. The *event header* shows the strike, dip, rake, centroid depth and scalar seismic moment (in Nm) of the minimum misfit solution. *Top* focal sphere shows the lower hemisphere stereographic projection of the P waveform nodal planes, and the positions of the seismic stations used in the modelling routine. *Lower* focal sphere shows the SH nodal planes. *Capital letters* next to the station codes correspond to the position on the focal sphere. These are ordered clockwise by azimuth, starting at north. *Solid lines* are the observed waveforms, *dashed lines* are the synthetics. The inversion window is marked by *vertical lines* on each waveform. The source time function (STF) is shown, along with the time scale for the waveforms. The amplitude scales for the waveforms are shown below each focal sphere. The P and T axes within the P waveform focal sphere are shown by a *solid* and an *open circle*, respectively

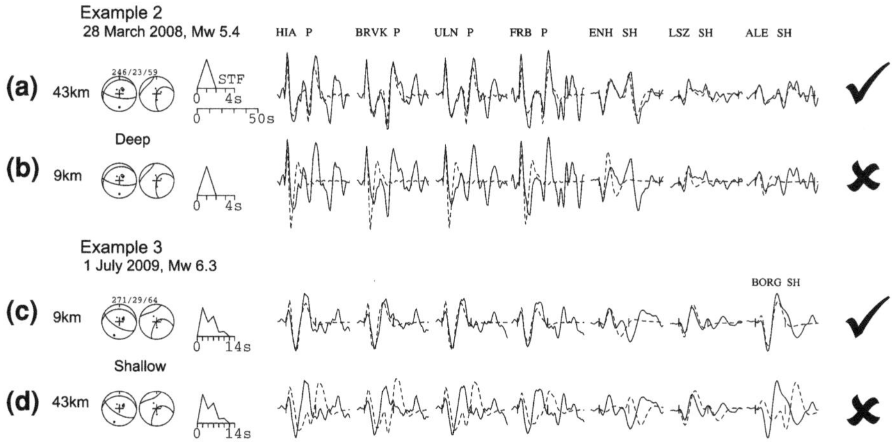

Fig. 3.3 P and SH waveforms recorded at seven stations (4P, 3SH) for two earthquakes that occurred within about 20 km of each other on 28 March 2008 (*Example 2*) and 1 July 2009 (*Example 3*). Both involved thrust faulting, but one occurred at 43 km depth, along the subduction zone interface, whilst the other occurred at a depth of 9 km. Synthetics are calculated for both events at a depth of 9 (**b, c**) and 43 km (**a, d**). The fit of observed (*solid*) to synthetic (*dashed*) waveforms is good in (**a**) and (**c**) and poor in (**b**) and (**d**), confirming their different depths

In this area, Bohnhoff et al. [52] use wide-angle seismic data to infer an interface at 20–25 km, which they interpret to be the subduction zone interface in this region. However I suspect that this earthquake at 43 $\pm$ 4 km depth occurred on the subduction zone interface because: (a) it has a shallowly dipping thrust mechanism; (b) it has a slip vector that is parallel to the GPS-measured motion between the Aegean and Nubia in this region; (c) it occurs at almost exactly the same depth as other interface events to the SW of Crete (which are in a similar position relative to the main bathymetric escarpment of the Hellenic trench system). The alternative is that this earthquake occurred in the downgoing Nubian lithosphere but, as I show later in Sect. 3.4, nearly all other earthquakes in the downgoing lithosphere have P-axes that are arc-parallel (showing contraction along the arc), whereas this earthquake has a P-axis that is perpendicular to the arc.

Another thrust-faulting earthquake occurred in the same part of Hellenic subduction zone on 7 July 2009, about 20 km east of the interface earthquake described above (Example 3, Fig. 3.1). This earthquake, of M_w 6.4, has a shallow centroid depth (9 km) and a plane which dips to the north at 30° (slightly steeper than the 20–25° dip of the subduction zone interface in this area). Seismograms at selected stations are matched much better by synthetics calculated at 9 km depth than by a centroid depth of 43 km (Fig. 3.3c, d), which is the depth of the inferred subduction interface earthquake in the same region (Example 2), suggesting that this thrust occurred along a splay in the overriding Aegean material. The mechanisms and depths of these two earthquakes, in almost the same place, support the hypothesis that seismogenic splay faults branch from the main subduction zone interface.

3.4 The Subducting Africa-Nubia Lithosphere

As the Africa (Nubia) lithosphere subducts to the north beneath the Aegean, earthquakes within the slab reveal its position to a depth of about 160 km (Fig. 3.4, [53, 54]). With a down-dip length of 300 km and a present-day convergence rate of 35 mm/year between the Aegean and Africa, a seismogenic slab of this length could have been produced in 8.5 Ma, if the convergence rate has been constant. This is less than the observed 10–12 Ma that is typically needed for slabs to heat up enough to become aseismic (e.g. [55, 56]), and raises the possibility that the subduction zone itself may only be 8.5 Ma old. But Africa-Eurasia convergence has occurred since at least the Cretaceous [57] and most of the convergence rate in the Hellenic subduction zone arises from the rapid extension in the Aegean Sea and anticlockwise rotation of Turkey, which may be relatively recent at its present rate (e.g. [15, 58, 59]). A long episode of relatively slow subduction prior to about 3–5 Ma may be responsible for a possible asesimic high-velocity prolongation of the slab penetrating much deeper (to at least 600 km) than the deepest earthquakes [60] and widespread andesitic volcanism in the northern Aegean of early Miocene age [61]. The geometry of the seismogenic part of the subducting slab has been reviewed by Hatzfeld and Martin [53, 54] and Papazachos et al. [16], who show that it is relatively flat beneath the Peloponnese, steepening up beneath the Gulf of Corinth, whereas it is more steeply dipping (50°) in the east, near Rhodes (Fig. 3.4b).

Figure 3.4 shows the earthquakes whose depths and locations make it likely they occurred within the subducting lithosphere. Although their focal mechanisms are quite varied, they in fact reveal a relatively simple pattern. Taymaz et al. [12] noticed that, south of Crete, the mechanisms of these earthquakes generally had P axes parallel to the local strike of the subduction zone; a pattern seen to be more widespread in the arc by Benetatos et al. [18]. In the new enlarged dataset it is now clear (Fig. 3.4c) that this pattern is indeed general round most of the arc, both for earthquakes up to 200 km seaward of the main bathymetric scarp of the Hellenic Trench and to depths of 100 km within the subducting slab itself. There is no evidence for a change in the pattern as the depth increases, and thus no evidence that the slab undergoes a dramatic change in stress state at 50–60 km as claimed by Ganas and Parsons [31].

The T axes of these earthquakes are, in general, radial to the arc and aligned down-dip within the subducting slab (Fig. 3.4c, [18]). In some respects this pattern is typical of other subduction zones where the seismically-active slabs do not penetrate as deep as 670 km [62, 63], where earthquakes shallower than 410 km generally have down-dip T axes. The usual explanation given for this pattern is that the slab is lengthening under its negative buoyancy unless its tip is supported by penetrating to the density contrast at 670 km. In this case it suggests that, even if an older warm and aseismic continuation of the Hellenic slab exists far beyond the deepest earthquakes, it has insufficient strength to influence the earthquake mechanisms at shallow depth.

The along-strike shortening of the downgoing African lithosphere may be related to the strong curvature of the Hellenic subduction zone and its subducting slab. In particular, the consistency of the along-strike P axes and down-dip T axes, even as

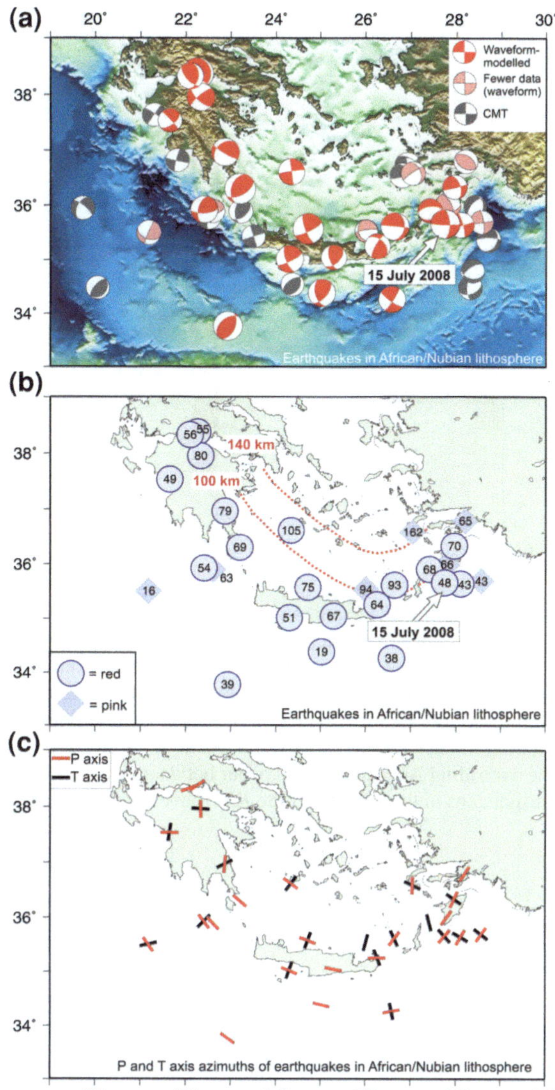

Fig. 3.4 Earthquakes in the down-going Nubian lithosphere. **a** Focal mechanisms are ranked by quality in the order *red*, *pink*, *grey* (see key and text). The majority are either strike-slip or thrust, and reveal a broad trend of along-arc compression. The location of the event of 15 July 2008, discussed as Example 1 in Fig. 3.2 is indicated. **b** Well-constrained depths, in kilometres, of waveform-modelled earthquakes [*red* and *pink* in (**a**)]. Depths increase to the north within the subducting slab, which dips more steeply in the east than in the west. *Dashed red lines* are contours of the slab depth taken from Hatzfeld and Martin [53]. **c** Azimuths of the P (*red*) and T axes (*black*), with a plunge of less than 45°, of earthquakes in (**b**) show a regular concentric pattern of along-arc shortening, both seaward of the Hellenic Trench and within the subducting slab. T axes are generally radial to the arc and down-dip in the descending slab

the slab forms itself into a conical surface that increases its curvature with depth, suggests that there is a change in Gaussian curvature that may be related to the seismicity. A surface that changes its Gaussian curvature must deform by changing its surface area, which may account for some of the seismicity distribution within the slab. In the case of the Hellenic subduction zone there are too few earthquakes to define the slab geometry over its entire depth range with sufficient precision to test this hypothesis in the manner attempted for the Tonga slab by Nothard et al. [64].

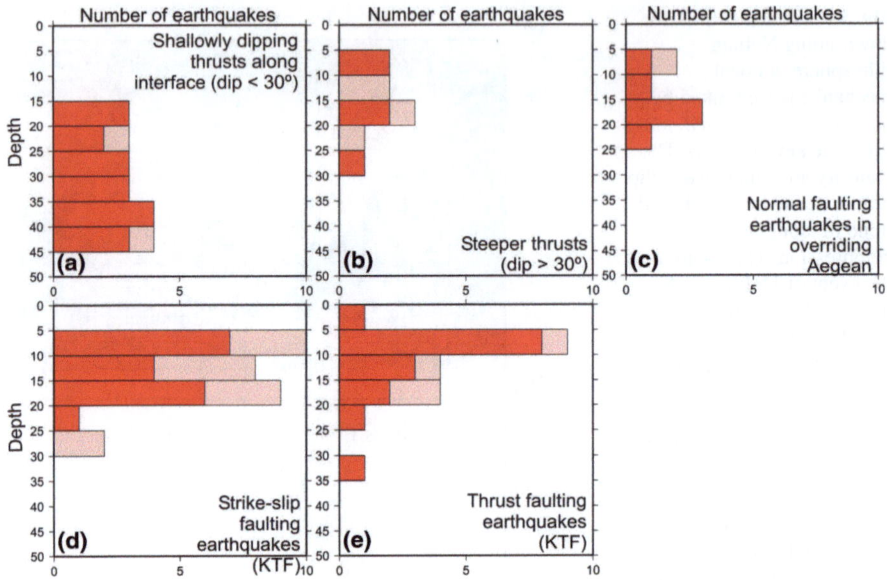

Fig. 3.5 Earthquake depth ranges (in kilometres) in different tectonic settings. *Solid red bars* show earthquakes for which there is abundant good quality waveform-modelled data. *Pink bars* represent earthquakes whose depths are well-resolved from waveform modelling, but for which there are insufficient data for the focal mechanism to be determined independently. **a** Shallow-dipping earthquakes along the subduction interface occur at depths between 15 and 45 km. **b** More steeply dipping thrust-faulting earthquakes occur at depths of 5–30 km. **c** Normal-faulting earthquakes in the overriding Aegean material occur between 5 and 25 km. **d, e** Both strike-slip and thrust-faulting earthquakes in the vicinity of the KTF have similar centroid depths (0–35 km)

3.5 Africa-Aegean Convergence

Although earthquakes in the subduction zone account for less than 10% of the convergence between Africa and the Aegean (Sect. 3.2.1), there are sufficient moderate-sized events of M_w 5.0–6.0 to reveal the geometry and depth of the subduction interface in the depth range ~15–45 km. There are few earthquakes shallower than 15 km (Fig. 3.5a), probably because of the thick unconsolidated sediment on the down-going plate and, in common with subduction zones elsewhere, the interface becomes aseismic deeper than ~45 km because of thermal or dewatering effects [65–67]. In these respects, the Hellenic subduction zone displays a typical distribution of inter-plate seismicity.

Figure 3.6 shows the distribution of thrust and reverse-faulting earthquakes on the subduction zone interface and immediately above it. The epicentres lie in a relatively compact band 50–100 km wide, roughly following the trace of the bathymetric scarps of the Hellenic Trench. The up-dip continuation of the interface beneath the accretionary prism of the Mediterranean ridge is apparently aseismic. The along-strike distribution of seismicity round the arc is patchy, with dense groups of earthquakes separated by areas without any earthquakes large enough for long period body-wave modelling. This irregular distribution is also seen in microearthquake studies [37, 38]. It is unclear whether this patchiness is a long-term feature of the subduction zone, or merely reflects the limited time of observation. It is, however, noteworthy that earthquakes seem to have concentrated in the same areas over three decades; the clusters in Figs. 3.6 and 3.7 do not represent one mainshock and its aftershocks. This may indicate the presence of persistent long-term rough, or well-coupled, frictional patches along the interface (but with relatively small surface areas), surrounded by other areas of stable sliding or creep. The most prominent patches are south of Crete beneath Gavdos, beneath the SW corner of Crete, and SW of the Peloponnese.

Low-angle thrust-faulting earthquakes on the subduction interface show some variation in depth with position around the arc (Fig. 3.6b). In the east, near Rhodes, the only available fault-plane solutions are from first-motion studies, with no precise depth constraints. In this region hypocentral depths from the EHB catalogue, while unlikely to be reliable to better than $\sim\pm$ 15 km [40, 42], are mostly between 20 and 45 km, which is comparable with other parts of the arc. To the south of Crete, well-determined earthquake depths on the interface occur as deep as 40–45 km, with some shallower events at $\sim$20 km beneath Gavdos. Beneath the SW corner of Crete, well-determined interface earthquakes are at a depth of 40 km. In western Crete the African Moho, imaged by receiver function studies, is at a depth of 55 km [68], approximately 10 km below the depth of the seismogenic interface, suggesting that the interface follows the top of the subducting oceanic crust. The next substantial patch of earthquakes is south of the Peloponnese, where well-determined depths for low-angle thrusts are typically at 25–30 km. Further west, beneath the Ionian Islands (discussed later), they are shallower still, at 15–20 km. The trend seems to be for the maximum depth of the seismogenic part of the interface to decrease towards the NW and this trend seems to correlate broadly with a decrease in sediment thickness in the same direction [5, 8, 69]: where the sediment is thickest the seismogenic part of the interface is deeper, and vice versa.

The mechanisms of the majority of these earthquakes are similar, although the dips of the landward-dipping low-angle fault planes vary between 10° and 20°. For these events, the steep south-dipping auxiliary nodal plane, and therefore the slip-vector, is well-constrained as it passes through the centre of the teleseismic station distribution on the focal sphere. The dip of the fault plane itself is less well-constrained, but there is a suggestion of a consistent spatial variation, with the shallowest dips (5–10°) in the west of the arc, and steeper dips (approaching 25°) to the south of Crete. The shallower dips in turn are associated with thinner sediment and a shallower depth of interface events.

In addition to shallow-dipping thrust earthquakes along the interface, several reverse-faulting earthquakes with steeper dips (>30°) generally occur at shallower depths than the interface (Fig. 3.5b) within the overriding material (Fig. 3.6c). These events presumably occur on splay faults that merge with the interface at depth and accommodate some of the convergence in the subduction zone. Their centroids are typically $<$ 20 km depth, whereas low-angle thrusts on the interface in the same

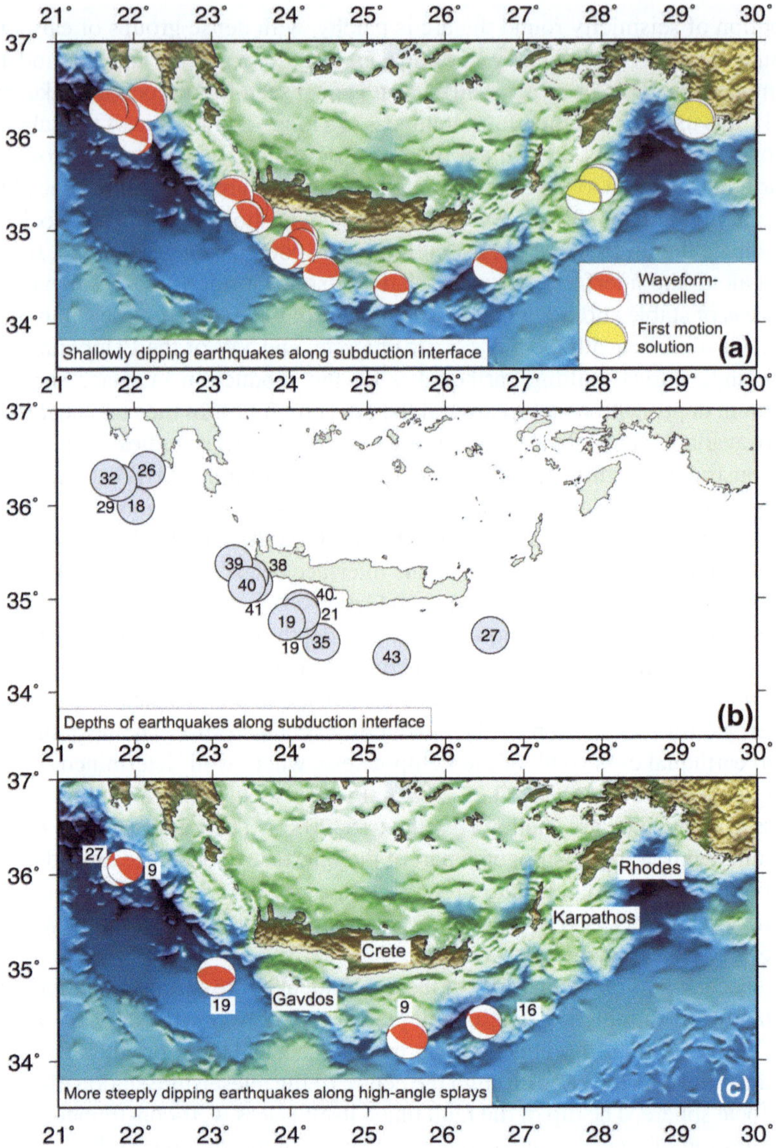

Fig. 3.6 Reliable waveform-modelled fault plane solutions (*red*) and selected first-motion solutions (*yellow*) related to Nubia-Aegean convergence. **a** Shallow-dipping thrust faulting earthquakes which occur along the plate interface; **b** their depths in kilometres. **c** More steeply dipping fault plane solutions which generally occur at shallower depths than the interface and probably represent slip along splay faults

regions are deeper, reaching a maximum depth of 40–45 km depth (Fig. 3.6c). It was one such splay fault beneath SW Crete which I suggested in Chap. 2 was responsible for the catastrophic AD 365 earthquake and tsunami.

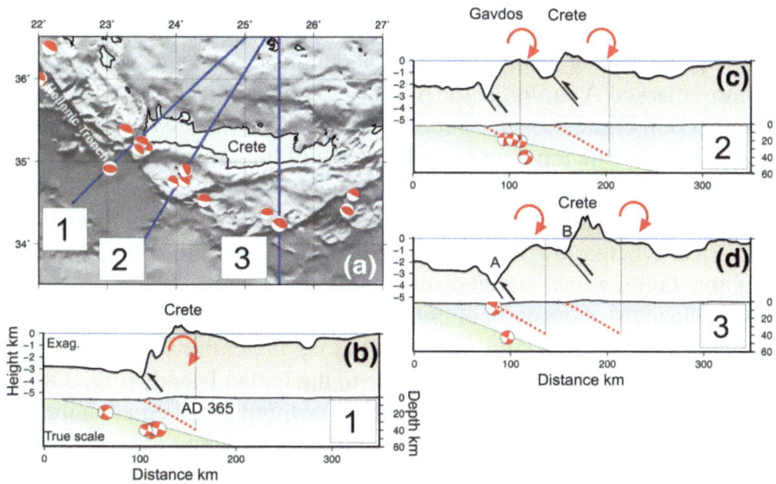

Fig. 3.7 Geometry of faults that accommodate the main Nubia-Aegean shortening around the Hellenic arc. **a** The three cross-sections (**b-d**) pass through clusters of earthquakes which are plotted as *red fault plane solutions* and are taken from the same subset, with good depth control, as is plotted in Fig. 3.6. **b** This cross-section passes through the SW corner of Crete, which was uplifted in the AD 365 earthquake. The *lower section* is true-scale, whereas the *upper section* is highly exaggerated in the vertical. A possible geometry for the fault that caused the uplift is plotted as a *red dashed line*. This fault projects to the surface at the Hellenic Trench and seems to control both the southern and northern coasts of Crete: the northern coast lies directly above the deepest extent of the fault. **c** Possible fault geometry between Crete and Gavdos [symbols as in (**b**)]. **d** Cross-section in the region of Examples 2 and 3 [see Fig. 3.3 and the text; symbols as in (**b**)]

Figure 3.7 shows bathymetric and topographic cross-sections through the three main clusters of convergence-related earthquakes in the region of Crete. Each of the cross-sections is plotted so that it runs perpendicular to the local strike of the Hellenic Trench. The first cross-section passes through western Crete, where the subduction zone interface is at a depth of 40–45 km beneath the Hellenic Trench. In the interpretation in Chap. 2, the earthquake responsible for the 9 m of coastal uplift in AD 365 occurred along a splay fault which reaches the surface at the Hellenic Trench. In the model that best fits the uplift data, the fault extends to 45 km depth, and this bottom of the fault lies directly beneath the northern coast of Crete, suggesting that this splay may be responsible for the position and tilting of Crete itself.

The second cross-section passes through both Gavdos and Crete. There are two clear bathymetric escarpments, both of which may be the surface expressions of splay faults responsible for uplift and tilting. If splays dipping at 30° to a depth of 40 km (dotted red lines) are projected to the surface at the base of these steep escarpments, there is again a reasonable match between the deepest extent of the fault and length scale of tilting for the northern fault in this cross-section.

Cross-section 3 passes through the earthquakes described in Examples 2 and 3 in Figs. 3.1 and 3.3. I argue in Sect. 3.3.5 that the shallower earthquake of 1 July 2009 occurred on a splay fault in the hanging wall of the main subduction zone interface. The surface projection of its north-dipping nodal plane coincides with a clear bathymetric scarp, marked A in Fig. 3.7d. Another steep bathymetric escarpment occurs further north (B in Fig. 3.7d) which could be the surface expression of another north-dipping splay fault. The topography to the north of both these escarpments appears to be tilted N, again on a length scale consistent with the hanging wall of a fault with a 30° dip extending through the entire seismogenic thickness. In both sections 2 and 3, it is not clear whether the northern inferred splay faults are still active: it may be that the active faulting has migrated south with time.

It is possible to use the pattern of uplift from the AD 365 earthquake as a template to identify sites of possible long-term uplift along the entire straight section of the Hellenic Arc from the south of central Crete to the Ionian Islands (Fig. 3.8) if points an equal distance from the Hellenic trench are uplifted by an equal amount. If this entire section of the arc behaves in the same way as that suggested for western Crete in Chap. 2, then both the Mani and Methoni peninsulas, and possibly Kythera, fall within the expected area of long-term uplift. The Mani peninsula has well-recorded Plio-Quaternary terraces [70, 71], and the Methoni peninsula is another obvious place to look for evidence of long-term uplift related to the subduction. Unlike in western Crete, there is no clear uplifted Holocene shoreline on either of these peninsulas, perhaps suggesting that if they are uplifted in very large earthquakes, such an event has not occurred since sea-level stabilised about 6 ky ago; an argument discussed further in Chap. 2.

The slip-vector azimuths of the thrust and reverse faulting earthquakes in Fig. 3.6 are shown in Fig. 3.1b. Along most of the arc, their direction closely resembles the GPS velocity vectors for the overriding Aegean material relative to Nubia (Fig. 3.9). In the eastern part of the arc, this similarity in azimuth implies that although the relative motion of the plates is oblique to the bathymetric trenches in this region, the inter-plate motion is taken up by oblique slip along the subduction zone interface (discussed further in Sect. 3.7). South-west of Crete and the Peloponnese, the match in direction between velocity vectors and slip-vectors is particularly close.

The only significant deviation from this simple picture occurs in the far west of the arc, near the KTF (longitude 20–22° E), where the slip vectors of the thrust earthquakes are almost perpendicular to the GPS vectors (as can be seen in Fig. 3.13b). This region is discussed separately in Sect. 3.8.

Both the GPS velocities and the slip-vectors of earthquakes on the subduction zone interface show a consistent change in azimuth from west to east of about 20°. The divergence in the earthquake slip vectors would require a change in length of about 15 mm/year along the arc, assuming a radial convergence of 35 mm/year between Nubia and the southern Aegean in the Hellenic Trench system (as shown in Figs. 3.1b, 3.9) which must be taken up by either extension in the overriding material or shortening in the underthrusting material (or both). The GPS measurements of Hollenstein et al. [50] show 19 mm/year extension around the arc, implying that the majority of the divergence seen in the earthquake slip vectors is taken up by extension in the overriding material. This extension is revealed by the earthquakes discussed in the next section.

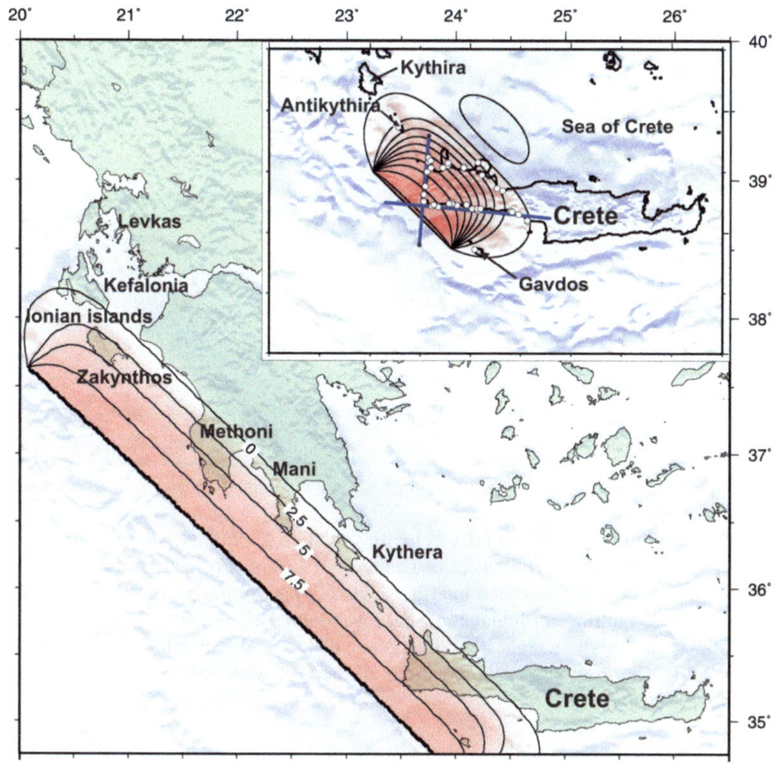

Fig. 3.8 *Inset* shows the inferred uplift distribution during the AD 365 earthquake on Crete from Chap. 2 based on an elastic dislocation model and constrained by observations at the *circled* locations. Contours are at metre intervals, with a maximum of 9 m uplift. The *main figure* shows this uplift distribution extrapolated along strike, with contours of uplift at intervals of 2.5 m. If this section of the Hellenic Trench experiences earthquakes similar to that of AD 365, then this pattern may be used to predict areas that may be uplifted in the long-term, such as the Mani and Methoni peninsulas and the island of Kythera. The maximum uplift is the same in the *main figure* and the *inset*

3.6 East-West Extension in the Overriding Aegean

A third important group of earthquakes occurs within the overriding Aegean and is not directly concerned with accommodating Africa-Aegean convergence. These earthquakes represent E-W extension (Fig. 3.10), mostly on N-S normal faults, and are generally shallower than 20 km (Fig. 3.5c). Their orientations contrast with the generally E-W striking normal faulting events that characterise the N-S extension of the central and northern Aegean, Western Turkey and mainland Greece [10, 72].

Although Fig. 3.10 shows N-S normal faulting earthquakes mainly east and NW of Crete and in the Peloponnese, normal faults of this trend can be recognised in the topography and bathymetry all round the arc [11, 73–76]. It is likely that the

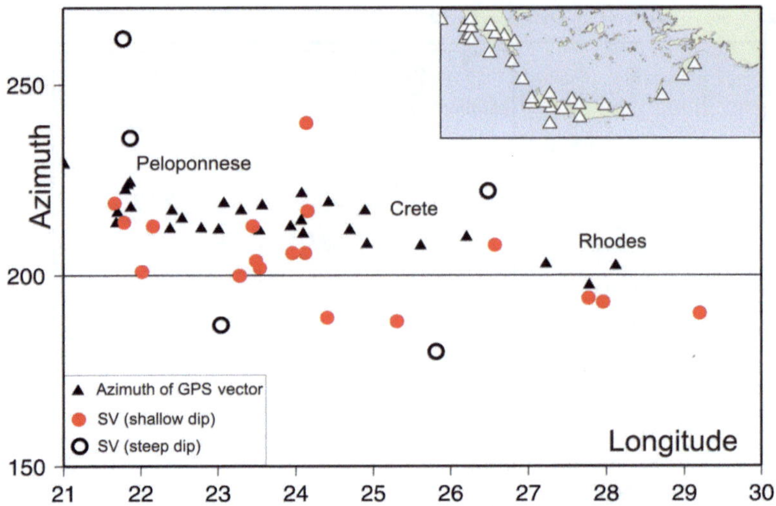

Fig. 3.9 Variation in the azimuths of GPS velocities relative to Nubia (*triangles*) and earthquake slip-vectors (*circles*) around the Hellenic arc. The *red circles* are slip vectors of shallowly-dipping earthquakes along the subduction-zone interface; *open circles* represent the slip-vectors of more steeply dipping thrust-faulting earthquakes at shallower depths in the overriding material. The GPS vectors and slip-vectors show a consistent divergence of just under 20 mm/year between the west and east of the arc. The *inset* shows the positions of the GPS stations whose velocity vectors are plotted in the *main panel*

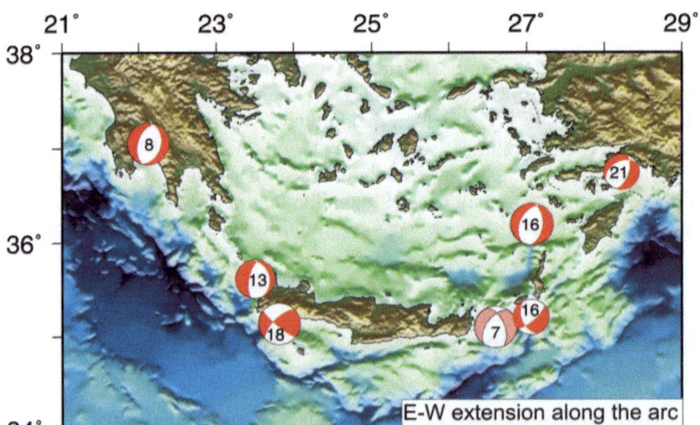

Fig. 3.10 Earthquakes indicating E–W extension in the overriding Aegean material above the subduction-zone interface. Centroid depths (in kilometres) are plotted in the focal mechanisms. The highest-quality waveform-modelled solutions are in *red*; those in *pink* have also been modelled, but with fewer data. The E–W extension represented by these earthquakes has been concentrated in the east and west of the arc, rather than in the centre, over the period of observation

continental crust between the Peloponnese and Crete is below sea-level because it has been stretched to a greater extent than the adjacent landmasses. From recent earthquakes, it appears that extension is concentrated in the east and west of the arc, rather than in the centre. The GPS vectors along the arc show the largest divergence in azimuth between 25° E and 29° E longitude, supporting the observation that extension is occurring relatively rapidly in the east of the arc. Several authors [28, 77] have suggested that the rapid Quaternary uplift and northward tilting of Crete is caused by the underplating of sediments from the accretionary prism, as they are underthrust beneath Crete on high-angle reverse faults splaying off the subduction interface. As Pirazzoli et al. [78], Stiros and Drakos [27] and Shaw et al. [28] point out, it is the uplift of Crete that has preserved the evidence which allowed the causative faulting of the AD 365 earthquake to be inferred. If uplift associated with similar underplating between Crete and the Peloponnese has been removed by crustal thinning associated with N-S normal faulting, then so too has any evidence of very large earthquakes of the AD 365 type also been removed from this section of the subduction zone. The potential for such large earthquakes in this part of the zone is consequently not known.

Thus, while it is clear the N-S striking normal-faulting earthquakes seen in western Crete and the Peloponnese by Lyon-Caen et al. [73], Armijo et al. [74] and Taymaz et al. [12] continue to the eastern end of the arc, there are no new earthquakes in the west to add to those earlier studies.

3.7 The Eastern Termination

The active tectonics of the eastern end of the subduction zone remains obscure. There are few earthquakes compared with other parts of the arc; some of them have well-constrained focal mechanisms and depths, but many are either within the Nubian lithosphere or related to the E-W extension of the overriding Aegean. In Fig. 3.11 I show the earthquakes whose mechanisms and depths suggest they could reflect some part of the primary Nubia-Aegean motion across the zone. The earthquakes do not line up to reveal a major fault, but do indicate two main types of faulting:

(a) The first are thrust-faulting earthquakes with a shallow northward dip (Fig. 3.11a), whose slip vectors closely parallel the GPS vectors. They are capable of taking up the motion between Nubia and the Aegean on the subduction zone interface. These earthquakes may be associated with hummocky E-W ridges perpendicular to their slip vectors (Fig. 3.11a, c).
(b) The second type are compatible with NE-SW left-lateral strike-slip faulting (Fig. 3.11b), with slip vectors in the range SSW to WSW. The strike of the fault-planes and of their slip-vectors are sometimes parallel to local bathymetric lineations (Fig. 3.11c) which may represent multiple strike-slip faults.

A type of faulting that is conspicuously absent is thrusting with a strike parallel to the NE-SW deforming zone and an orthogonal (i.e. SE) slip vector. Thus there is no

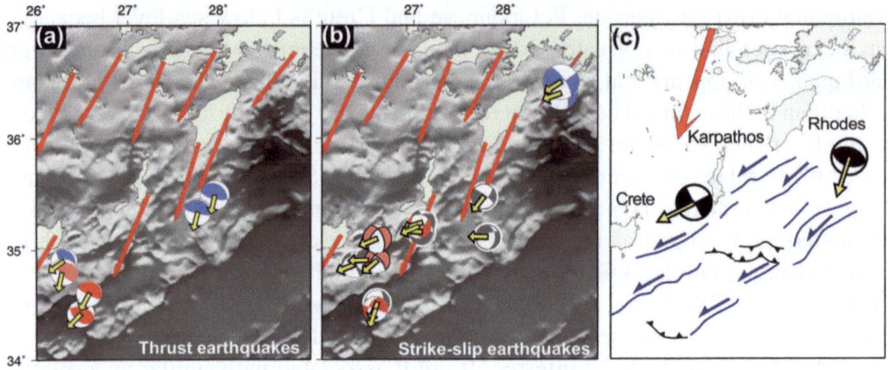

Fig. 3.11 Earthquakes at the eastern end of the Hellenic arc that contribute to the main Nubia-Aegean convergence. *Focal spheres* are coloured by quality as in Fig. 3.4, with the addition of first-motion solutions, which are plotted in *blue*. **a** Thrust faulting earthquakes have slip vectors (*yellow*) sub-parallel to the GPS velocities (*red*). **b** Strike-slip faulting earthquakes have slip vectors oblique to the GPS velocities, but their strikes align with bathymetric lineations. **c** An interpretive summary, showing strike-slip fault strands as *blue lines* and the hummocky surface expression of blind thrusts as *black lines*

evidence for the classic form of slip partitioning common in subduction zones [79, 80]. Instead, the whole region is distinctly broken-up, with hummocky topography that may represent the surface expression of thrust faults and some more linear features that may represent strike-slip faults. This deformation is distributed in the overriding material, which must be underlain by a relatively shallow north-dipping interface, as seismicity in the Nubian slab continues to a depth of over 100 km (Fig. 3.4b, [53]). The distributed strike-slip faulting, with slip vectors oblique to those on the thrusts, is a style of deformation familiar from Mongolia [81] and Iran [82], which can contribute to the NNE-SSW convergence if the strike-slip faults rotate clockwise with time. In Mongolia and Iran the rotating strike-slip faults often terminate in thrusts: south of Rhodes and Karpathos, this may be the origin of the hummocky ridges. The analogous behaviour in an extensional setting is seen in central Greece, where strike-slip faults end in extensional graben [83].

3.8 The Western Termination

3.8.1 Strike-Slip Faulting Earthquakes

An abrupt bathymetric escarpment, around 4 km deep, and sometimes called the KTF [84], runs along the NW edge of the Ionian Islands, and forms the western termination of the Hellenic subduction zone. It is associated with a high level of seismicity, much of it compatible with NE-SW right-lateral strike-slip faulting [85].

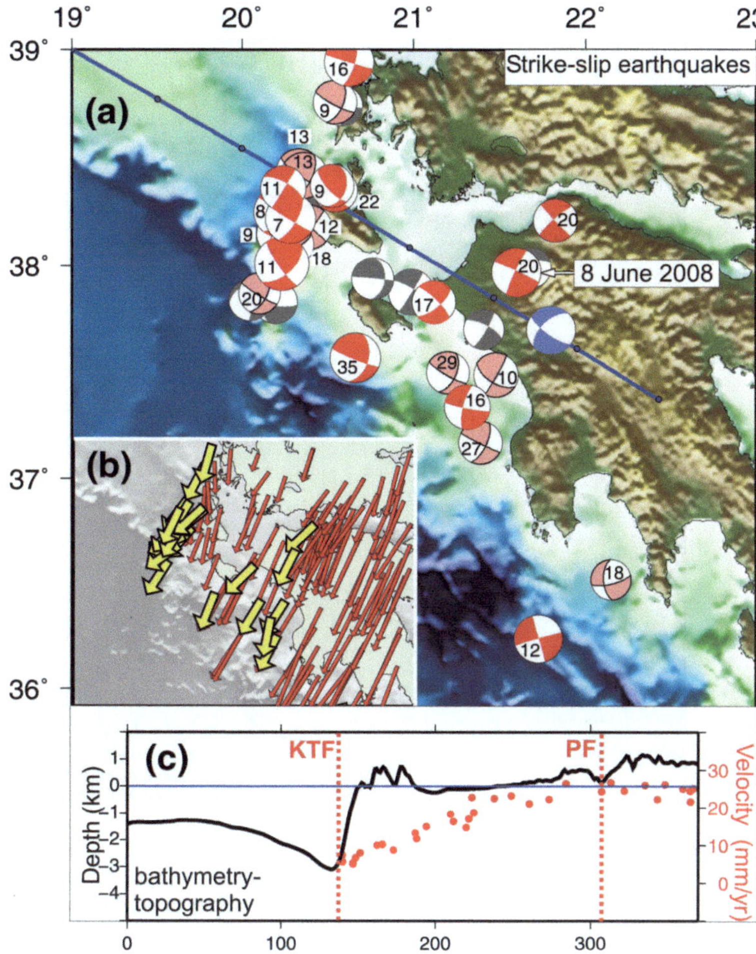

Fig. 3.12 Strike-slip faulting earthquakes at the western termination of the subduction zone. **a** Well-determined depths are given as *numbers* adjacent to the focal mechanisms. The colour of the focal mechanism denotes its method of determination and the quality of the data, as in Figs 3.4 and 3.11. The *blue line* running from NW to SE is the line of cross-section in (**c**). **b** Slip vectors of the waveform-modelled earthquakes (*yellow arrows*) and the GPS velocities relative to Nubia (*red arrows*). **c** Bathymetry-topography (*black line*) along the line of section, alongside the trenchperpendicular component of the GPS velocities (*red dots*). The magnitude of the GPS velocities gradually increases between the KTF and the PF, revealing a region of distributed shear that is consistent with the spread of earthquake focal mechanisms

Louvari et al. [19] suggested it consisted of two segments with different strikes: a southern segment 90 km long that strikes N-E, and a northern segment 40 km long striking NNE. There is some support for these two different segments in the different slip vectors north and south of Kefalonia in Fig. 3.12b.

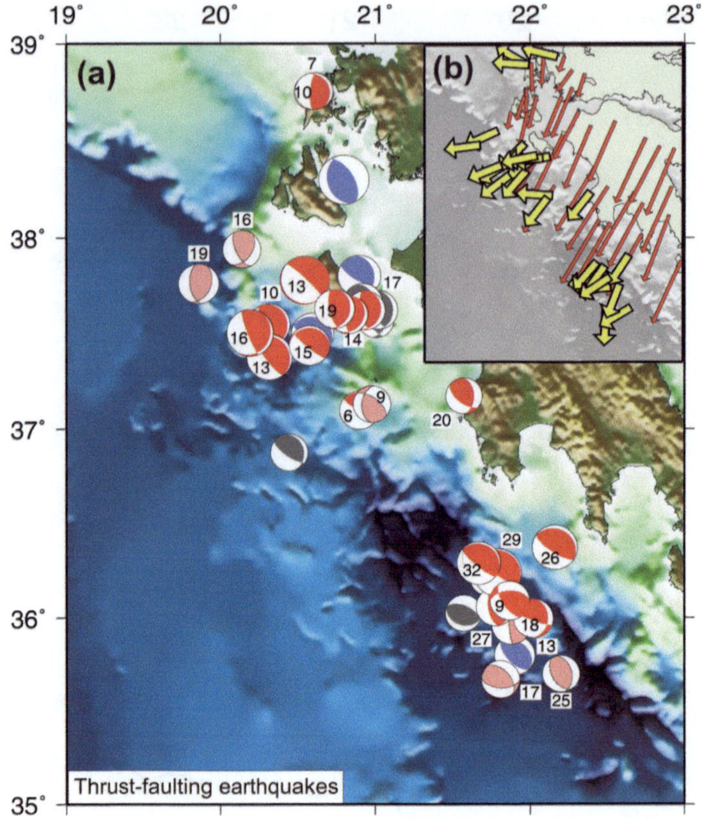

Fig. 3.13 Thrust faulting earthquakes at the western termination of the subduction zone. Well-determined depths are given as numbers adjacent to the focal mechanisms. The colour of the focal mechanism denotes its method of determination and the quality of the data, as in Fig. 3.12. The *inset* shows the slip vectors of the waveform-modelled earthquakes (*yellow*) and the GPS velocities relative to Nubia (*red*)

In addition to earthquakes along the two segments of the KTF, Louvari et al. [19] recognised a few strike-slip earthquakes which did not occur along this prominent bathymetric escarpment, but were further to the SE. A recent strike-slip earthquake of M_w 6.3 on 8 June 2008 in the NW Peloponnese, close to the city of Patras (Fig. 3.12), belongs to this group. Its aftershocks defined a zone running NE-SW along a line separating mountainous topography to the SE from the gently undulating, broad coastal plain to the NW [86, 87]. The focal mechanism is well-determined by waveform modelling and indicates NE-SW right-lateral strike-slip faulting with an unusually deep centroid at 20 km. From its seismic moment, the estimated source dimension is of order 10–15 km, explaining the lack of a continuous co-seismic surface rupture.

The NE-SW right-lateral shear that crosses central Greece [3] and is seen clearly in the GPS velocity field (Fig. 3.1b, [4]) is accommodated mostly by clockwise rotations of normal faults and the blocks they bound about a vertical axis [83, 88, 89]. At the western termination of the subduction zone, it is instead accommodated by NE-SW right-lateral strike-slip faulting, which includes the KTF but is also distributed over a zone $\sim$100 km wide to the SE. The slip-vectors of strike-slip faulting earthquakes are almost parallel to the GPS vectors in this region (Fig. 3.12b). The distributed strike-slip motion is seen both in the focal mechanisms (Fig. 3.12a) and the steady gradient in GPS velocities (Fig. 3.12b). Figure 3.12c is a cross-section through the topography and bathymetry in a direction perpendicular to the KTF (blue line). The magnitude of the SW component of the GPS velocities is plotted in red, and reveals a gradual SE-ward increase in SW-directed velocity from nearly zero to the NW of Kefalonia, to $\sim$30 mm/year, close to the total Aegean-Nubian convergence, to the south-east of the Patras Fault (PF), recognised from GPS and seismicity. It is clear that there is not a single dextral strike-slip fault, but rather several strands accommodating the motion, similar to the situation in the North Aegean, where the North Anatolian fault also branches into several strands [49].

3.8.2 *Thrust Faulting Earthquakes*

To the south of the Peloponnese, the thrust faulting earthquakes are typical of those interface events seen elsewhere around the arc (Fig. 3.13). The earthquakes along the interface occur at depths of 20–25 km and are on shallowly dipping thrust faults. However beneath the Ionian Islands (Kefalonia, Zakynthos and Levkas) the thrust faulting earthquakes have a different character: their strike is N-S and their slip vectors are almost perpendicular to the GPS convergence direction (Fig. 3.13b), suggesting a degree of slip partitioning in this region.

Figure 3.5d and e show histograms of the depths of strike-slip and thrusting earthquakes near the western termination. Here the subduction zone interface is relatively shallow, so almost all of the earthquakes are shallower than 20 km depth. Both the thrust and strike-slip earthquakes have a similar depth distribution.

3.8.3 *Gravity and Flexure*

North-west of the KTF, the bathymetry deepens and the negative free-air gravity anomaly increases towards the SE, resembling a flexural signal. Using the technique described in detail by McKenzie [90], NE-SW gravity and topography profiles were stacked within the box outlined in red in Fig. 3.14 and the resultant profiles matched by the flexure of an elastic beam with thickness T_e by D. McKenzie. Consistent values for T_e of 7–8 km ($\pm$1.5 km) were obtained either from the gravity or the topography (Fig. 3.14) though the plots of misfit against T_e are broad and do not have well-determined minima. These values of T_e are similar to those estimated for Italy using the admittance method of gravity and topography by D'Agostino and McKenzie [91]. Such low values are typical of young continental regions [92] and suggest that the area NW of Kefalonia, i.e. the S Adriatic, is indeed continental and that the change in water depths and earthquake centroid depths north of 40° N marks the transition from the subduction of oceanic lithosphere to the south to continent-continent collision in NW Greece.

The KTF juxtaposes high-topography (Ionian Islands) with the deeper sea floor of the Ionian Sea. Much of this relief is probably inherited by the SW motion of the Ionian Islands on the KTF and some may be enhanced by the thrusting of the Ionian Islands to the west causing the Ionian Sea to flex in response (Fig. 3.15). The GPS velocity vectors do not show any significant movement of the Ionian Islands normal to the KTF, but the islands would act as a load regardless of how they were emplaced (i.e. whatever the relative direction of motion between the Ionian Islands and the Southern Adriatic). The agreement between GPS velocities and slip vectors on the strike-slip faults shows that the loading is principally by emplacement along a lateral ramp, perpendicular to the main thrust front in the SW. The flexural loading is presumably more obvious at the western termination of the arc (i.e. west of Kefalonia) than south of the Peloponnese because of the thick sediments in the south. The N-S striking thrust faulting earthquakes may be related to the collapse of the overthrust material in the east along the steep bathymetric escarpment of the KTF.

3.9 The Connection with the Aegean and Central Greece

The abundant data from earthquake focal mechanisms, slip vectors and GPS velocities reveal how faulting accommodates the large-scale motions of the Aegean and Eastern Mediterranean. The most obvious features of the GPS velocity field are the apparent flow of Aegean material towards the SW (Fig. 3.1b), the relative rigidity of the southern Aegean and the shear between Europe and Turkey, which continues through the North Aegean into central Greece [50, 93, 94]. Earthquakes in the Eastern Mediterranean closely reflect this velocity field: shallowly dipping thrust earthquakes along the plate interface reveal the convergence between the Aegean and Nubia; the southern Aegean is largely aseismic [58]; and strike-slip earthquakes align in strands through the North Aegean [83].

Where the NE-SW right-lateral strike-slip fault branches in the north Aegean reach the eastern coast of Central Greece, they connect to the rapidly opening gulfs of Volos and Evia [83], which are opening fastest in the east [95–97]. Yet the most seismically active and rapidly opening graben system in this region is the Gulf of Corinth, which opens fastest at its western end, where it appears to connect with the strike-slip fault system on which the 2008 Patras earthquake occurred. Likewise, the KTF projects to the western end of the Gulf of Arta which appears to be another normal-faulted basin, like other graben on the west coast, such as the Trichonis basin [83, 98]. As the slip vectors on all the E-W normal faults are roughly N-S [83] whereas the overall motions

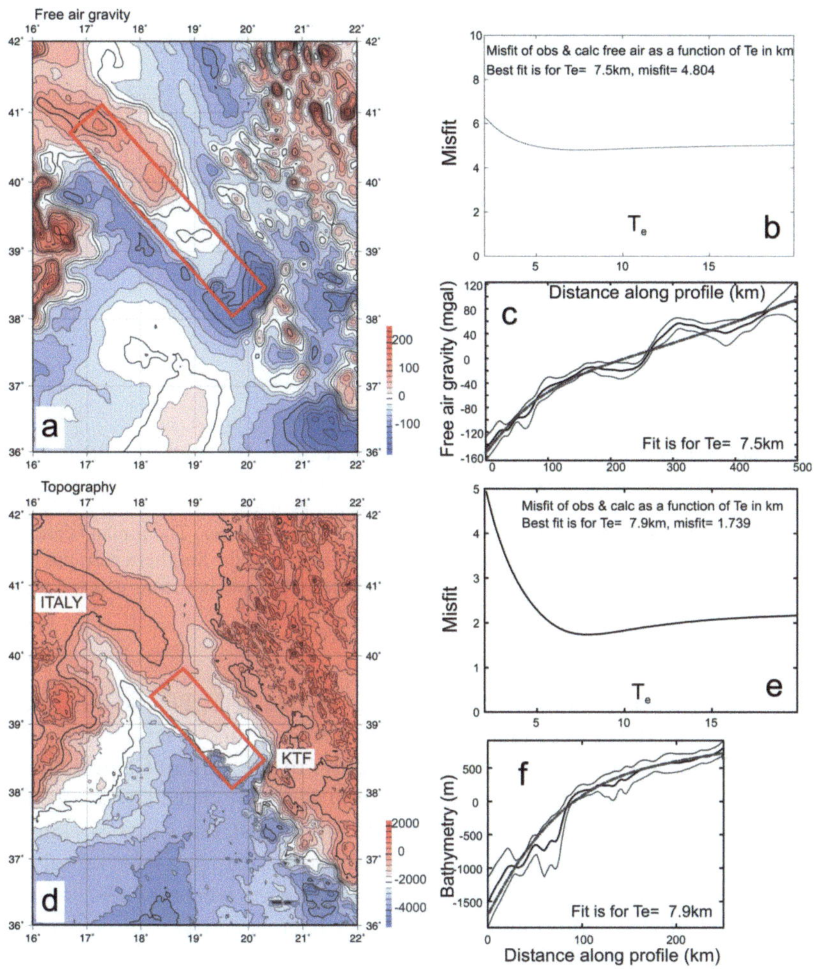

Fig. 3.14 **a** Free-air gravity map for the Southern Adriatic [100]. Profiles within the *red box* were stacked and then modelled by a flexed elastic beam to estimate the effective elastic thickness (T_e) by McKenzie, following the procedure of McKenzie [90]. **b** The resulting misfit between observed and calculated free air gravity is plotted against T_e , showing a broad, poorly-resolved minimum, with a formal best fit for a T_e value of 7.5 km. **c** Stacked gravity profiles against distance along the profile (in kilometres), with a 1σ error envelope. The *thick dashed line* shows the best-fitting solution. **d** Map of topography [99] in the region. Profiles within the *red box* were stacked and then also modelled as a flexed beam to estimate T_e . **e** Resulting misfit between observed and calculated topography as a function of T_e . In this case the minimum, though broad, is clearer and the best-fitting solution for T_e is 7.9 km, consistent with the result obtained from the gravity modelling. **f** Stacked profiles against distance along the profile (in kilometres), with a 1σ error envelope. The *thick dashed line* shows the best-fitting solution

are to the SW, the blocks between the normal faults must rotate about vertical axes [88]. It is this block rotation, confirmed in places by paleomagnetic measurements [89], that accommodates shear in central Greece. Thus the connection to strike-slip

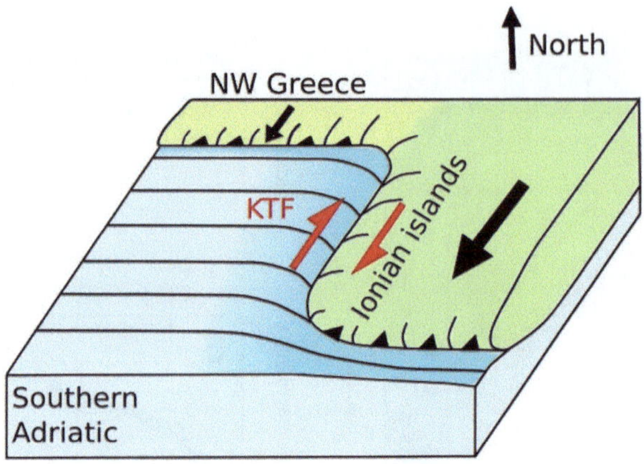

Fig. 3.15 Interpretation of the kinematics in the area of the KTF. The Southern Adriatic is underthrust beneath NW Greece along a line of shallowly-dipping thrust faulting earthquakes that run along the coastline [85, 101]. The Ionian Islands and the Peloponnese are thrust to the SW, at a rate of up to 35 mm/year, overriding the oceanic lithosphere of the Ionian Sea. The Ionian Islands act as a load on the Southern Adriatic, emplaced along a lateral ramp, and the Southern Adriatic flexes in response. Shear between the Ionian Islands and the Southern Adriatic is accommodated along the KTF

faults on the western side of Central Greece, along the KTF, PF and other parallel faults in some ways resembles that on the eastern side of this system, in the Aegean. Figure 3.16 is a cartoon showing my interpretation of the region. Central Greece is sheared in a right-lateral sense. The whole region accommodates a gradient in SW-directed velocity, relative to the north (Fig. 3.1). In the east, this shear is taken up by strike-slip faulting which ends in normal-faulted graben that open most rapidly in the east. In the west, another set of graben open fastest at their western ends, connecting to a second set of strike-slip faults in the Ionian Islands and W. Peloponnese. The areas between the normal faults rotate clockwise relative to the north, which is why the slip-vectors on the normal faults are N-S, not NE-SW. Thus the graben of Central Greece act as a regional relay zone, connecting two strike-slip faulting systems in the NE and SW. In reality, the blocks are not entirely rigid and there are many minor faults at the ends of the major faults. The simplified cartoon of Fig. 3.16 considers only the major active faults in the region and only their relation to the present-day earthquakes and GPS motions.

Figure 3.17 is a cartoon summarising the main tectonic features of the Hellenic subduction zone between NW Greece and Central Crete, integrating the insights of Goldsworthy et al. [83] in the northern part of the area with those of this study. The African slab subducts to the North, undergoing along-arc compression as it descends in to the arcuate subduction zone, accompanied by along-arc extension in the overriding Aegean. In the cross-section, Santorini marks the position of the volcanic arc (shown as a triangle in Fig. 3.17). The Nubian lithosphere is covered in a thick pile of sediments and the Aegean overrides these sediments along splay

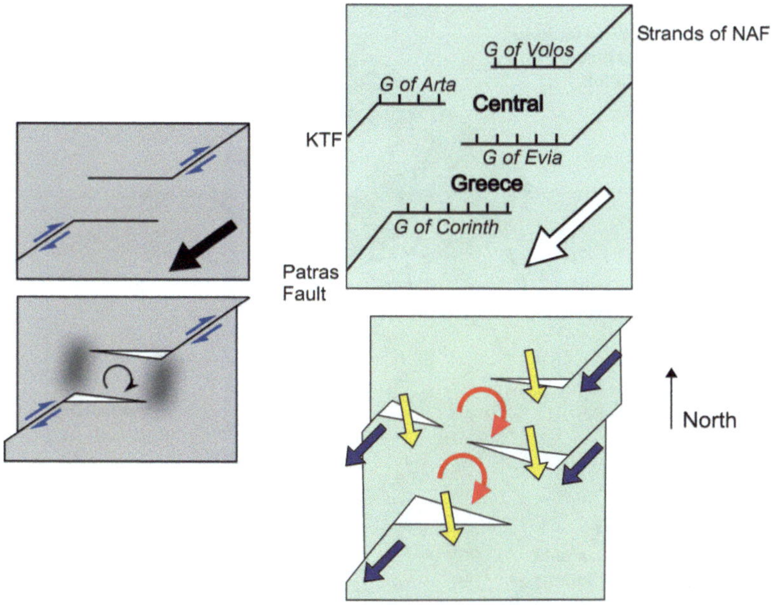

Fig. 3.16 Cartoon showing my interpretation of the kinematics of Central Greece, extended from Goldsworthy et al. [83]. For locations, see Fig. 3.1. On the *left* is a simple case, in which right-lateral shear is transferred through a region by extensional faulting and block rotation. When motion occurs along the strike-slip faults, the central block rotates and diffuse deformation is expected at the edges of this block (shaded region in *bottom panel*). The *panels on the right* show how this simple concept can be applied to Central Greece. Rather than a single strike-slip fault offset by a normal faulting system, parallel strike-slip faults connect with graben and the blocks between these graben rotate. In the east, several strands of the North Anatolian Fault (NAF) connect to graben which open most rapidly at their eastern ends (Volos and Evia). In the west, the graben open most rapidly at their western ends and connect to the KTF and PF. This panel shows how strike-slip faults terminate in normal-faulted graben, which accommodate the shear by rotations (*red arrows*). The strike-slip faulting earthquakes have slip vectors (*blue*) which are parallel to the direction of shearing, whereas the slip-vectors of normal-faulting earthquakes are N–S. The blocks between the normal-faulted graben will rotate clockwise. Note that the structural expression of rotations is in graben whose extension varies rapidly along strike

faults, leading to uplift of Crete. Shear between the Aegean and Eurasia is taken up by strike-slip faulting in the Aegean and in the region of the Ionian Islands, but is accommodated by block rotation and normal faulting in Central Greece.

3.10 Conclusions

The aim of this chapter was to use earthquake source parameters and GPS data to elucidate kinematic features of the Hellenic subduction zone. The principal new results of this study are:

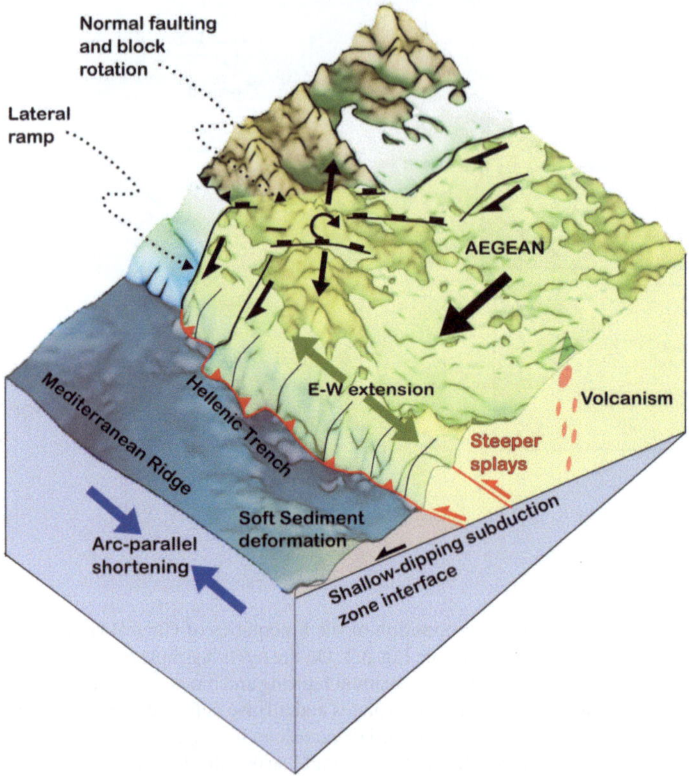

Fig. 3.17 Cartoon summarising the main tectonic features of the Hellenic subduction zone, between NW Greece and Crete. The descending Nubian lithosphere is shown in *blue*, overlain by a thick accretionary prism (*grey*) which forms the Mediterranean Ridge. As the Nubian lithosphere enters the arcuate subduction zone, it contracts in an arc-parallel direction. The overriding Aegean is shown as *yellow* and deforms by arc-parallel extension along N–S normal faults. The Hellenic Trench marks the position of a splay fault, along which Aegean material overthrusts the accretionary prism, causing uplift. The Aegean is moving SW with respect to Eurasia and this relative motion is taken up by strike-slip faults in the Aegean and Ionian Islands. Between these strike-slip faulting systems, Central Greece deforms by normal faulting and block rotation, acting as a regional relay zone. Apulia is flexed beneath the Ionian islands due to a lateral ramp (see Fig. 3.15)

1. I have shown that careful re-evaluation of earthquake focal mechanisms and depths, and their comparison with the now excellent GPS velocity data, allows the seismicity and deformation within the subducting Nubian lithosphere, the overriding Aegean lithosphere, and on the interface between them, to be distinguished; a result previously demonstrated only near Crete [12].
2. I confirm that deformation nearly everywhere within the Nubian lithosphere is characterised by concentric along-arc shortening, mostly by reverse and strike-slip faulting, and show that it is seen in earthquakes as far as 200 km seaward of

Crete and to depths of at least 100 km north of Crete. The ubiquity of this pattern suggests it is related to the strong along-strike curvature of the subduction zone.

3. Earthquakes on the subduction interface itself are low-angle thrusts in the depth range 15–45 km, generally reaching a maximum depth of 20 km in the west and 45 km in the centre of the arc, near Crete. Steeper-dipping reverse faults occur above the main interface in most parts of the arc, and probably occur on splay faults connecting with the subduction interface at depth.
4. With the depth extent and dip of the seismogenic part of the main subduction interface well-defined, I confirm that, even allowing for the contribution of small earthquakes and the occasional very large AD 365-type event, the main subduction zone interface must be largely uncoupled and aseismic. The principal large earthquake and tsunami hazard is likely to be from rare AD 365-type events on splay faults, as described in Chap. 2.
5. The close agreement between GPS velocities relative to Nubia and earthquake slip vectors on the subduction interface, both of which are radial to the arc in the centre and west, requires an along-arc change in length. Most of this is probably achieved by along-strike extension of the overriding arc.
6. In the west, the subduction zone terminates in a steep bathymetric escarpment that is predominantly a strike-slip fault (KTF; e.g. [85]), which also marks a change from oceanic subduction to continent–continent collision. Earthquake mechanisms, GPS data and a flexural gravity anomaly show that the origin of the escarpment is as a lateral ramp, formed as the Ionian Islands are emplaced SW onto the Apulian lithosphere. Some thrust faulting with slip vectors perpendicular to the ramp reveal an element of slip partitioning in the region, which enhances the relief of the scarp.
7. GPS data and strike-slip earthquake mechanisms show that NE-SW right-lateral shear at the western end of the subduction zone is not restricted to the KTF, but is distributed on several parallel strike-slip faults over a distance of ~150 km. One of these was responsible for the destructive earthquake of 8 June 2008 near Patras.
8. At the eastern end of the subduction zone low-angle thrust faults with slip vectors oblique to the strike of the arc, but parallel to the Nubia-Aegean convergence measured by GPS, occur on the main subduction zone interface. The interface is overridden by material that is pervasively deformed by distributed strike-slip faults responsible for linear escarpments, and hummocky bathymetry that may be associated with blind splay thrusts. Slip vectors on the strike-slip faults are oblique to those on the thrusts and to the Nubia-Aegean convergence, suggesting that the strike-slip faults rotate about vertical axes.
9. The GPS velocity field in Greece shows a 150–200 km wide band of NE-SW right-lateral shear extending from the NE Aegean to the SW termination of the Hellenic subduction zone (e.g. [88, 93]). In the NE and SW this motion is achieved by parallel NE-SW strike-slip faults. In central Greece it is achieved by clockwise-rotating blocks bounded by E-W normal faults with N-S slip vectors. In both the east and west of central Greece, the strike-slip faults terminate in graben that die out towards the centre of Greece. The rapid gradients of

extension along-strike of these graben is the structural manifestation of the clockwise rotation. Central Greece thus acts as a relay zone connecting the strike-slip faults to either side.

References

1. E. Chaumillon, J. Mascle, Variations latérales des fronts de déformation de la Ride Méditeranéene. Bulletin de la Société Géologique de France **166**, 463–478 (1995)
2. E. Chaumillon, J. Mascle, From foreland to forearc domains: new multichannel seismic reflection survey of the Mediterranean Ridge accretionary complex (Eastern Mediterranean). Marine Geol. **138**, 237–259 (1997)
3. D. McKenzie, Active tectonics of the Mediterranean region. Geophys. J. R. Astron. Soc. **30**, 109–185 (1972)
4. R. Reilinger, S. McClusky, P. Vernant, S. Lawrence, S. Ergintav, R. Cakmak, H. Ozener, F. Kadirov, I. Guliev, R. Stepanyan, M. Nadariya, G. Hahubia, S. Mahmoud, K. Sakr, A. ArRajehi, D. Paradissis, A. Al-Aydrus, M. Prilepin, T. Guseva, E. Evren, A. Dmitrotsa, S.V Filikov, F. Gomez, R. Al-Ghazzi, G. Karam, GPS constraints on continental deformation in the Africa-Arabia–Eurasia continental collisional zone and implications for the dynamics of plate interactions. J. Geophys. Res. **111**, B05411 (2006)
5. X. Le Pichon, J. Angelier, J. Aubouin, N. Lyberis, S. Monto, V. Renard, H. Got, K. Hsu, Y. Mart, J. Mascle, D. Matthews, D. Mitropoulos, P. Tsoflias, G. Chronis, From subduction to transform motion: a seabeam survey of the Hellenic trench system. Earth Planet. Sci. Lett. **44**, 441–450 (1979)
6. X. Le Pichon, Landlocked oceanic basins and continental collision: the Eastern Mediterranean as a case example, in *Mountain Building Processes* (Academic Press, London, 1982), pp. 201-211.
7. N.H. Kenyon, R.H. Belderson, A.H. Stride, Detailed tectonic trends on the central part of the Hellenic outer ridge and in the Hellenic trench system. Geol. Soc. Lond. **10**, 335–343 (1982)
8. K.A. Kastens, A.B. Nancy, M.B. Cita, Progressive deformation of an evaporite-bearing accretionary complex: Sea-MARCI, SeaBeam and piston core observations from the Mediterranean Ridge. Marine Geophys. Res. **14**, 249–298 (1992)
9. J.P. Foucher, N. Chamot-Roocke, S. Alexandry, J.M. Augustin, S. Monti, P. Pavlakis, M. Voisset, in *Multibeam Bathymetry and Seabed Reflectivity Maps of the MEDRIFF Corridor Across the Eastern Mediterranean Ridge*, ed. by Terra Cognita, EUG (VII) (Blackwell, London, 1993), pp. 278–279
10. D. McKenzie, Active tectonics of the Alpine–Himalayan belt: the Aegean Sea and surrounding regions. Geophys. J. Int. **55**, 217–254 (1978)
11. C. Kreemer, N. Chamot-Rooke, Contemporary kinematics of the southern Aegean and the Mediterranean Ridge. Geophys. J. Int. **157**, 1377-1392 (2004)
12. T. Taymaz, J. Jackson, R. Westaway, Earthquake mechanisms in the Hellenic Trench near Crete. Geophys. J. Int. **102**, 695–731 (1990)
13. A. Camerlenghi, M.B. Cita, W. Hieke, T. Ricchiuto, Geological evidence for mud diapirism on the Mediterranean Ridge accretionary complex. Earth Planet. Sci. Lett. **109**, 493–504 (1992)
14. J. Mascle, E. Chaumillon, An overview of Mediterranean Ridge collisional accretionary complex as deduced from multichannel seismic data. Geo-Marine Letters, **18**, 81–89 (1998)
15. J. Angelier, N. Lyberis, X. Le Pichon, P. Huchon, The tectonic development of the Hellenic Trench and Sea of Crete: a synthesis. Tectonophysics **86**, 159–196 (1982)
16. B.C. Papazachos, V.G. Karakostas, C.B. Papazachos, E.M. Scordilis, The geometry of the Wadati-Benioff zone and lithospheric kinematics in the Hellenic Arc. Tectonophysics **319**, 275–300 (2000)

17. A. Kiratzi, E. Louvari, Focal mechanisms of shallow earthquakes in the Aegean Sea and the surrounding lands determined by waveform modelling: a new database. J. Geodyn. **36**(1-2), 251–274 (2003) Active faults: analysis, processes and monitoring
18. C. Benetatos, A. Kiratzi, C. Papazachos, G. Karakaisis, Focal mechanisms of shallow and intermediate depth earthquakes along the Hellenic Arc. J. Geodyn. **37**, 253–296 (2004)
19. E. Louvari, A.A. Kiratzi, B.C. Papazachos, The Cephalonia Transform Fault and its extension to western Lefkada Island (Greece). Tectonophysics **308**, 223–236 (1999)
20. J. Jackson, D. McKenzie, The relationship between plate motions and seismic moment tensors, and rates of active deformation in the Mediterranean and Middle East. Geophys. J. **93**, 45–73 (1988)
21. N.N. Ambraseys, C.P. Melville, R.D. Adams, *The Seismicity of Egypt, Arabia and the Red Sea: A Historical Review* (Cambridge University Press, Cambridge, 1995)
22. E. Guidoboni, A. Comastri, Catalogue of Earthquakes and Tsunamis in the Mediterranean Area from the 11th to the 15th Century. (Istituto Nazionale di Geofisica e Vulcanologia, Rome, 2005)
23. N.N Ambraseys, *Earthquakes in the Mediterranean and Middle East* (Cambridge University Press, Cambridge, 2009)
24. N.N. Ambraseys, Data for the investigation of the seismic sea-waves in the Eastern Mediterranean. Bull. Seismol. Soc. Am. **52**, 895–913 (1962)
25. B.C. Papazachos, Large seismic faults in the Hellenic Arc. Ann. Geophys. **39**, 892–903 (1996)
26. E. Guidoboni, A. Comastri, The large earthquake of 8 August 1303 in Crete: seismic scenario and tsunami in the Mediterranean. J. Seismol. **1**, 55–72 (1997)
27. S. Stiros, A. Drakos, A fault model for the tsunami-associated, magnitude > 8.5 eastern Mediterranean, AD 365 Earthquake. Zeitschrift für Geomorphologie **146**, 125–137 (2006)
28. B. Shaw, N.N. Ambraseys, P.C. England, M.A. Floyd, G.J. Gorman, T.F.G. Higham, J.A. Jackson, J.-M. Nocquet, C.C. Pain, M.D. Piggott, Eastern Mediterranean tectonics and tsunami hazard inferred from the AD 365 earthquake. Nature Geosci. **1**, 268–276 (2008)
29. H. Kanamori, Seismic and aseismic slip along subduction zones and their tectonic implications, in *Island Arcs, Deep Sea Trenches and Back-Arc Basins*, ed. by M. Talwani, W.C. Pitman III, Maurice Ewing Series I, Amer. Geophys. Union 163–174 (1977)
30. L. Ruff, H. Kanamori, Seismic coupling and uncoupling at subduction zones. Tectonophysics **99**, 99–117 (1983)
31. A. Ganas, T. Parsons, Three-dimensional model of Hellenic arc deformation and origin of the Cretan uplift. J. Geophys. Res. **114**, (2009). doi:10.1029/2008JB005599
32. P.A. Cowie, C.H. Scholz, Displacement-length scaling relationship for faults: data synthesis and discussion. J. Struct. Geol. **14**, 1149–1156 (1992)
33. E.R. Engdahl, R.D. Vander Hilst, R.P. Buland, Global teleseismic earthquake relocation with improved travel times and procedures for depth determination. Bull. Seismol. Soc. Am. **88**, 722–743 (1998)
34. N.N. Ambraseys, The relocation estimated of epicentres in Iran. Geophys. J. R. Astron. Soc. **53**, 117–121 (1978)
35. M. Berberian, Evaluatiuon of the instrumental and relocated epicentres of Iranian earthquakes. Geophys. J. R. Astron. Soc. **58**, 625–630 (1979)
36. D. Hatzfeld, G. Pedotti, P. Hatzidimitriou, K. Makropoulos, The strain pattern in the western Hellenic arc deduced from a microearthquake survey. Geophys. J. Int. **101**, 181–202 (1990)
37. D. Hatzfeld, M. Besnard, K. Makropoulosw, P. Hatzidimitriou, Microearthquake seismicity and fault-plane solutions in the southern Aegean and its geodynamic implications. Geophys. J. Int. **115**, 799–818 (1993)
38. T. Meier, M. Rische, B. Endrun, A. Vafidis, H.P. Harjes, Seismicity of the Hellenic subduction zone in the area of western and central Crete observed by temporary local seismic networks. Tectonophysics **383**, 149–169 (2004)
39. CMT (2009) Global cmt project: http://globalcmt.org

40. A. Maggi, J.A. Jackson, K. Priestley, C. Baker, A re-assessment of focal depth distributions in southern Iran, the Tien Shan and northern India: do earthquakes really occur in the continental mantle? Geophys. J. Int. **143**, 629–661 (2000)
41. T. Talebian, J. Jackson, A reappraisal of earthquake focal mechanisms and active shortening in the Zagros mountains of Iran. Geophys. J. Int. **156**, 506–526 (2004)
42. E.R. Engdahl, J.A. Jackson, S.C. Myers, E.A. Bergman, K. Priestley, Relocation and assessment of seismicity in the Iran region. Geophys. J. Int. **167**, 761–778 (2006)
43. P. Zwick, R. McCaffret, G. Abers, MT5 program, IASPEI Software Library 4 (1994)
44. R. McCaffrey, G. Abers, SYN3: a program for inversion of teleseismic body waveforms on microcomputers. Air Force Geophysics Laboratory Technical Report, AFGL-TR-88-0099, Hanscomb Air Force Base, MA. (1988)
45. R. McCaffrey, P. Zwick, G. Abers, SYN4 Program. IASPEI Softw. Libr. **3**, 81–166 (1991)
46. J.L. Nabelek, Determination of earthquake source parameters from inversion of body waves. Ph.D. thesis (MIT, MA, 1984)
47. R. McCaffrey, J. Nabelek, Earthquakes, gravity, and the origin of the Bali Basin: an example of a nascent continental fold-and-thrust belt. J. Geophys. Res. **92**, 441–460 (1987)
48. P. Molnar, H. Lyon-Caen, Fault plane solutions of earthquakes and active tectonics of the Tibetan Plateau and its margins. Geophys. J. Int. **99**, 123–153 (1989)
49. T. Taymaz, J. Jackson, D. McKenzie, Active tectonics of the north and central Aegean sea. Geophys. J. Int. **106**, 433–490 (1991)
50. C. Hollenstein, M. Mller, A. Geiger, H.-G. Kahle, Crustal motion and deformation in Greece from a decade of GPS measurements1993–2003. Tectonophysics **449**, 17–40 (2008)
51. Z. Altamimi, X. Collilieux, J. Legrand, B. Garayt, C. Boucher, ITRF 2005: a new release of the International Terrestrial Reference Frame based on time series of station positions and Earth orientation parameters. J. Geophys. Res. **112**, B09401 (2007)
52. M. Bohnhoff, J. Makris, D. Papanikolaou, G. Stavrakakis, Crustal investigation of the Hellenic subduction zone using wide aperture seismic data. Tectonophysics **343**, 239–262 (2001)
53. D. Hatzfeld, C. Martin, Intermediate depth seismicity in the Aegean defined by tele-seismic data. Earth Planet. Sci. Lett. **113**, 267–275 (1992)
54. D. Hatzfeld, On the shape of the subducting slab beneath the Peleponnese, Greece. Geophys. Res. Lett. **21**, 173–176 (1994)
55. R. Wortel, Seismicity and rheology of subducted slabs. Nature **296**, 553–556 (1982)
56. R. Wortel, Deep earthquakes and the thermal assimilation of subducting lithosphere. Geophys. Res. Lett. **13**, 34–37 (1986)
57. N. McQuarrie, J.M. Stock, C. Verdel, B.P. Wernicke, Cenozoic evolution of Neotethys and implications for the causes of plate motions. Geophys. Res. Lett. **30** (2003). doi:10.1029/2003GL017992
58. J. Jackson, Active tectonics of the Aegean region. Annu. Rev. Earth Planet. Sci. **22**, 239–271 (1994) Jackson 1994 is not cited in text.
59. R. Armijo, B. Meyer, A. Hubert, A. Barka, Westward propagation of the North Anatolian Fault into the northern Aegean: timing and kinematics. Geology **27**, 267–270 (1999)
60. W. Spakman, M.J.R. Wortel, N.J. Vlaar, The Hellenic subduction zone: a tomographic image and its geodynamic implications. Geophys. Res. Lett. **15**, 60–63 (1998)
61. G. Pe-Piper, D.J.W. Piper, Neogene backarc volcanism of the Aegean: new insights into the relationship between magmatism and tectonics. Geol. Soc. Am. Special Pap. **418**, 17–31 (2007)
62. B. Isacks, P. Molnar, Mantle earthquake mechanisms and the sinking of the lithosphere. Nature **223**, 1121–1124 (1969)
63. C. Frohlich, The nature of deep-focus earthquakes. Annu. Rev. Earth Planet. Sci. **17**, 227–254 (1989)
64. S. Nothard, D. McKenzie, J. Haines, J. Jackson, Gaussian curvature and the relationship between the shape and the deformation of the Tonga slab. Geophys. J. Int. **127**, 311–327 (1996)

65. C.H. Scholz, Earthquakes and friction laws. Nature **391**, 37–42 (1998)
66. R.D. Hyndman, M. Yamano, D.A. Oleskevich, The seismogenic zone of subduction thrust faults. Island Arcs. **6**, 244–260 (1997)
67. R.J. Stern, Subduction zones. Rev. Geophys. **40**, 3–1 (2002)
68. X. Li, G. Bock, A. Vafidis, R. Kind, H.-P. Harjes, W. Hanka, K. Wylegalla, van der M. Meijde, X. Yuan, Receiver function study of the Hellenic subduction zone: imaging crustal thickness variations and the oceanic Moho of the descending African lithosphere. Geophys. J. Int. **155**, 733–748 (2003)
69. X. Le Pichon, S.J. Lallemant, N. Chamot-Rooke, D. Lemeur, G. pascal, The Mediterranean Ridge backstop and the Hellenic nappes. Marine Geol. **186**, 111–125 (2002)
70. D. Kelletat, D. Gassert, Quartarmorphologische Untersuchungen im Kustenraum der Mani Halbinsel, Peleponnes. Zeitschrift für Geomorphologie N. F., Suppl **22**, 8–56 (1975)
71. A. Scheffers, D. Kelletat, A. Vott, S.M. May, S. Scheffers, Late Holocene tsunami traces on the western and southern coastlines of the Peloponnesus (Greece). Earth Planet. Sci. Lett. **269**, 271–279 (2008)
72. M. Goldsworthy, J. Jackson, Active normal fault evolution in Greece revealed by geomorphology and drainage patterns. J. Geol. Soc. **157**, 967–981 (2000)
73. H. Lyon-Caen, R. Gaulon, P. Papadimitriou, R. Armijo, J. Drakopoulos, N. Delibassis, V. Kouskouna, K. Makropoulos, J. Baskoutass, J. Latoussakis, D. Papananstassiou, G. Pedotti, The 1986 Kalamata (south Peloponnesus) earthquake: detailed study of a normal fault, evidences for east-west extension in the Hellenic arc. J. Geophys. Res. **93**, 14967–15000 (1988)
74. R. Armijo, H. Lyon-Caen, D. Papanastassiou, East-west extension and Holocene normal fault scarps in the Hellenic arc. Geology **20**, 491–494 (1992)
75. J.B. de Chabalier, H. Lyon-Caen, A. Zollo, A. Deschamps, P. Bernard, D. Hatzfeld, A detailed analysis of micro-earthquakes in western Crete from digital three-component seismograms. Geophys. J. Int. **110**, 347–360 (1992)
76. C. Huguen, J. Mascle, E. Chaumillon, J.M. Woodside, J. Benkhelil, A. Kopf, A. Volkonskaia, Deformational styles of the eastern Mediterranean ridge and surroundings from combined swath mapping and seismic reflection profiling. Tectonophysics **343**, 21–47 (2001)
77. X. Le Pichon, J. Angelier, The Aegean Sea. Philos. Trans. R. Soc. Lond. Ser A, Math. Sci. **300**, 357–372 (1981)
78. P.A. Pirazzoli, J. Laborel, S.C. Stiros, Earthquake clustering in the eastern Mediterranean during historical times. J. Geophys. Res. **101**, 6083–6097 (1996)
79. T.J. Fitch, Earthquake mechanisms in the Himalayan, Burmese, and Andaman regions and continental tectonics in Central Asia. J. Geophys. Res. **75**, 2699–2709 (1970)
80. R. McCaffrey, Slip partitioning at convergent plate boundaries of SE Asia. Geol. Soc. Lond. Special Publ. **106**, 3–18 (1996)
81. A. Bayasgalan, J. Jackson, J.-F. Ritz, S. Carretier, Field examples of strike-slip fault terminations in Mongolia and their tectonic significance. Tectonics **18**, 394–411 (1999)
82. R. Walker, J. Jackson, C. Baker, Surface expression of thrust faulting in eastern Iran: source parameters and surface deformation of the 1978 Tabas and 1968 Ferdows earthquake sequences. Geophys. J. Int. **152**, 749–765 (2003)
83. M. Goldsworthy, J. Jackson, J. Haines, Continuity of active fault systems in Greece. Geophys. J. Int. **148**, 596–618 (2002)
84. E. Kokinou, E. Papadimitriou, V. Karakostas, E. Kamberis, F. Vallianatos, The Kefalonia transform zone (Offshore Western Greece) with special emphasis to its prolongation towards the Ionian Abyssal Plain. Marine Geophys. Res. **27**, 241–252 (2006)
85. C. Baker, D. Hatzfeld, H. Lyon-Caen, E. Papadimitriou, A. Rigo, Earthquake mechanisms of the Adriatic sea and western Greece: implications for the oceanic subduction-continental collision transition. Geophys. J. Int. **131**, 599–594 (1997)
86. A. Ganas, E. Serpelloni, G. Drakatos, M. Kolligri, I. Adams, C. Tsimi, E. Batsi, On the M_w 6.4 SW-Achaia (western Greece) earthquake sequence of 8 June 2008: seismological, field, GPS observations and stress modeling. J. Earthq. Eng. **13**, 1101–1124 (2008)

87. K. Konstantinou, N. Melis, S.-J. Lee, C. Evangelidis, K. Boukouras, Rupture process and aftershocks relocation of the 8 June 2008 M_w 6.4 earthquake in northwest Peloponnese. western Greece. Bull. Seismol. Soc. Am. **99**, 3374–3389 (2009)
88. D. McKenzie, J. Jackson, The relationship between strain rates, crustal thickening, palaeomagnetism, finite strain and fault movements within a deforming zone. Earth Planet. Sci. Lett. **65**, 182–202 (1983)
89. M. Mattei, N. D'Agostino, I. Zananiri, D. Kondopoulou, S. Pavlides, V. Spatharas, Tectonic evolution of fault-bounded continental blocks: comparison of paleomagnetic and GPS data in the Corinth and Megara basins (Greece). J. Geophys. Res. **109**, B02106 (2004)
90. D. McKenzie, Estimating T_e in the presence of internal loads. J. Geophys. Res. **108** (2003). doi:10.1029/2002JB001766
91. N. D'Agostino, D. McKenzie, Convective support of long-wavelength topography in the Appennines (Italy). Terra Nova **11**, 228–233 (2002)
92. A. Maggi, J.A. Jackson, D. McKenzie, K. Priestley, Earthquake focal depths, effective elastic thickness, and the strength of the continental lithosphere. Geology **28**, 495–498 (2000)
93. S. McClusky, Global positioning system constraints on plate kinematics and dynamics in the eastern Mediterranean and Caucasus. J. Geophys. Res. **105**, 5695–5719 (2000)
94. R. Reilinger, S. McClusky, D. Paradissis, S. Ergintav, P. Vernant, Geodetic constraints on the tectonic evolution of the Aegean region and strain accumulation along the Hellenic subduction zone. Tectonophysics **488**, 22-30 (2010). doi:10.1016/j.tecto.2009.05.027
95. P.J. Clarke, R.R. Davies, P.C. England, B. Parsons, H. Billiris, P.A. Paradissis, P.H. Denys, V. Ashkenazi, R. Bingley, H.G. Kahle, M.V. Muller, P. Briole, Crustal strain in central Greece from repeated GPS measurements in the interval 1989–1997. Geophys. J. Int. **135**, 195–214 (1998)
96. P. Briole, A. Rigo, H. Lyon-Caen, J.C. Ruegg, K. Papazissi, C. Mitsakaki, A. Balodimou, G. Veis, D. Hatzfeld, A. Deschamps, Active deformation of the Corinth rift, Greece: results from repeated global positioning system surveys between 1990 and 1995. J. Geophys. Res. **105**, 25605–25625 (2000)
97. A. Avallone, P. Briole, A.M. Agatza-Balodimou, H. Billiris, O. Charade, C. Mitsakaki, A. Nercessian, K. Papazissi, D. Paradissis, G. Veis, Analysis of eleven years of deformation measured by GPS in the Corinth Rift Laboratory area. Comptes Rendus Geosci. **336**, 301–311 (2004)
98. A. Kiratzi, E. Sokos, A. Ganas, A. Tselentis, C. Benetatos, Z. Roumelioti, A. Serpetsidaki, G. Andriopoulos, O. Galanis, P. Petrou, The April 2007 earthquake swarm near Lake Trichonis and implications for the active tectonics in western Greece. Tectonophysics **452**, 51–65 (2008)
99. W.H.F. Smith, D.T. Sandwell, Global seafloor topography from satellite altimetry and ship depth soundings. Science **277**, 1957–1962 (1997)
100. D.T. Sandwell, W.H.F. Smith, Global marine gravity from re-tracked Geosat and ERS-1 altimetry: ridge segmentation versus spreading rate. J. Geophys. Res. **114**, B01411 (2009)
101. A. Copley, F. Boait, J. Hollingsworth, J. Jackson, D. McKenzie, Sub-parallel thrust and normal faulting in Albania and the roles of gravitational potential energy and rheology contrasts in mountain belts. J. Geophys. Res. **114**, B05407 (2009)

Chapter 4
Radiometric Dating of Uplifted Marine Fauna in Crete and Central Greece

4.1 Introduction

The study described in this chapter arose from the radiocarbon dating of uplifted marine fauna mentioned briefly in Chap. 2. In an attempt to confirm that the elevation of the uplifted shoreline in western Crete really represents deformation in a single earthquake, I obtained radiocarbon dates on uplifted marine organisms between the uplifted paleo-shoreline and modern sea-level. Corals [*Stenocyathus vermiformus, Caryophyllida* (sp.) and *Balanophyllia regia*] and bryozoans (*Myriapora truncata*) all gave radiocarbon ages consistent with uplift in a large historical earthquake in AD 365 (within errors, and after conversion from radiocarbon years to calendar years). Figure 4.1 shows examples of the coral and bryozoan samples collected. But 14 radiocarbon dates on uplifted *Lithophaga lithophaga*, a boring mollusc, were substantially older than AD 365 (by up to 2,000 years, even allowing for errors), and exhibited a much less consistent set of ages than the coral and bryozoan dates.

Although sea-level is rarely stable for long periods, its well-understood behaviour in the late Quaternary has allowed its use in countless studies of active tectonic processes and, in particular, its relative stability in the late Holocene has led to its widespread use as a datum in the investigation of past earthquakes in coastal regions (e.g. [1, 2]). These uses of sea-level as a datum are well illustrated in Greece, one of the fastest-deforming continental regions on Earth [3, 4], where uplift of the coastline is seen both in the footwalls of active normal faults in the Aegean Sea (e.g. [5, 6]) and in the hanging wall of the subduction zone between the Aegean and Mediterranean (e.g. [7–9]). The rates of uplift in these circumstances are estimated either by the correlation of uplifted marine terrace sequences with the Late Quaternary sea-level curve (e.g. [6, 10, 11]) or, on a more recent time-scale, by the direct dating of uplifted marine organisms by radiometric methods (e.g. [7, 9, 12, 13]). In this chapter, I investigate factors that limit the use and resolution of such radiometric dates, particularly for paleoseismological purposes (i.e. the association of particular uplift observations with particular historical earthquakes), and which have hitherto been little appreciated or understood.

B. Shaw, *Active Tectonics of the Hellenic Subduction Zone*, Springer Theses,
DOI: 10.1007/978-3-642-20804-1_4,

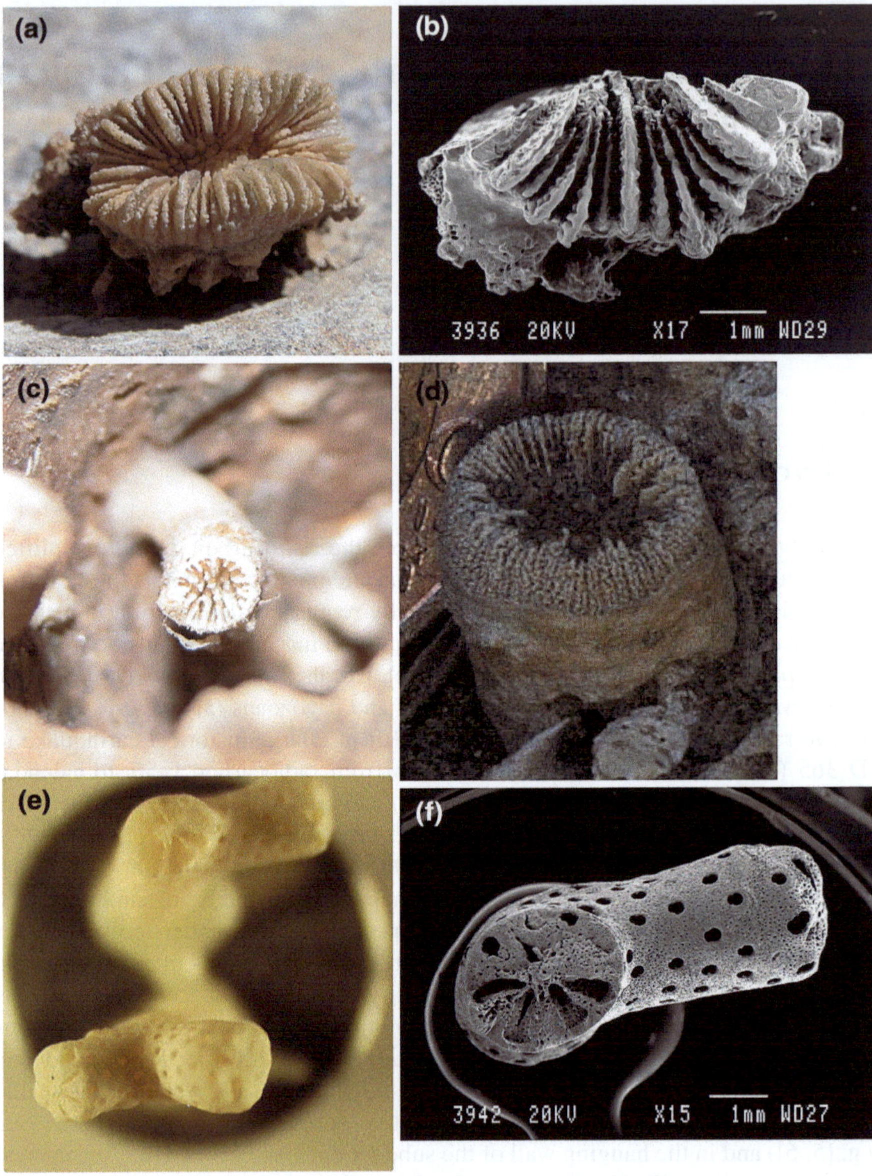

Fig. 4.1 Photographs of coral and bryozoan samples. **a, b** Images of *Balanophyllia regia*, **a** taken with a digital camera and **b** imaged using an SEM. The coral is approximately 1.5 cm in length. **c** Photograph of *Stenocyathus vermiformus* which was found growing within a vacated *L. lithophaga* hole. **d** Photograph of a specimen of *Caryophyllida* (sp.). This sample is 1 cm across. **e, f** Digital photograph and an SEM image respectively of the bryozoan *Myriapora truncata*

Lithophagids are widespread and abundant in the Mediterranean and Aegean Seas [14], and are frequently preserved in uplifted coastlines; much more so than corals and bryozoans. They are consequently much used to estimate uplift rates, and the discrepancy between their dates and those of corals and bryozoans is therefore a serious concern, raising the question of the limits to which their ages can be used to constrain uplift rates or the timing of particular uplift events.

Here, I demonstrate the importance of two effects for the interpretation of radiometric dates obtained for such marine organisms. The first, indicated by the sequence of biological colonisation of the marine substrate, is that the hard parts of organisms can remain within the substrate for long periods after death, so their ages may not date the uplift event itself. The second effect is the incorporation of host–rock carbon into the shells of the lithophagids, which causes radiocarbon ages to be artificially old. It is possible to quantify these effects using radiocarbon and U–Th dates on samples from Crete, and radiocarbon dates of museum specimens whose age of collection and death are known. While lithophagid ages remain a valuable means of estimating average uplift rates over thousands of years, the age offsets of up to 2,000 years that I demonstrate here make it unlikely that such ages can be used to reveal details of the uplift history. Finally, I illustrate these effects by considering Holocene rates of vertical movement in the extensional graben systems of the Gulfs of Corinth and Evia in central Greece.

4.2 Radiocarbon Dating

To compare critically radiocarbon dates of different organisms, and to associate them with specific dates, such as the date of an earthquake, a brief summary is needed of the procedures that are used in the dating process.

^{14}C is produced in the upper atmosphere by the interaction between ^{14}N and solar radiation, and then decays back to ^{14}N at a constant rate via beta decay. When an organism is alive, it is in equilibrium with the isotopic ratios of the medium in which it is growing (seawater for a coral). When it dies, it no longer exchanges carbon with its surrounding medium and it becomes a closed system in terms of carbon. The ^{14}C continues to decay at a constant rate, and the $^{14}C/^{12}C$ ratio falls predictably with time. If we know the original $^{14}C/^{12}C$ ratio of the seawater from which an organism grew, and the $^{14}C/^{12}C$ ratio of the shell today, we can calculate when the shell became a closed system, i.e. when the organism died [15, 16]. In principle, this method can determine the date of death of calcareous organisms up to about 50,000–60,000 years old, approximately ten times the half life of ^{14}C [15].

Various corrections are applied to account for other effects. All living organisms fractionate carbon, incorporating the lighter isotopes preferentially. The fractionation between ^{12}C and ^{13}C is half of that between ^{12}C and ^{14}C. The $^{12}C/^{13}C$ ratio is therefore also measured, and used to correct the measured $^{14}C/^{12}C$ ratio for the fractionation, known as the ‘vital effect’ [17, 18]. Since the correction is carried out using the isotopic ratios of the sample being measured, differences in fractionation between different organisms will not affect the dates obtained.

A 'conventional radiocarbon age', quoted in years before the present (BP: where the present is taken to be 1950), can then be determined. This is calculated (and defined) on the assumption that the atmospheric radiocarbon concentration has always been the same as it was in 1950, and that the half-life of radiocarbon is equal to the 'Libby' half-life of 5,568 years [19]. However, the 'Cambridge' half-life of 5,730 years [20] is thought to be more accurate. Additionally, the production of ^{14}C in the upper atmosphere varies over time as a result of changes in solar radiation flux and in the shielding ability of the Earth's magnetic field [16, 21, 22]. Both of these effects must be corrected for.

A further correction is required, when dating marine organisms, to allow valid comparison of results with samples in equilibrium with atmospheric and terrestrial carbon reservoirs. There is a delay in the exchange of carbon between the atmosphere and ocean, but the most important influence is the mixing of surface waters with ^{14}C-depleted carbon from the deep ocean [23]. The average difference between the radiocarbon age of a terrestrial sample, such as a tree, and that of a shell from the marine environment is about 400 radiocarbon years [24, 25], and is called a reservoir offset.

Radiocarbon ages for these samples were determined in the Research Laboratory for Archaeology and the History of Art (RLAHA) in Oxford. Using the INTCAL marine calibration dataset [26] and suitable software, it is possible to calibrate radiocarbon ages and thereby correct for the reservoir effect. A value of 53$\pm$ 43 year was calculated to account for the local reservoir offset from the modelled world ocean, using known-age data from the eastern Mediterranean [27, 28], and this was used in the calibration of these radiocarbon ages. The results of these analyses are displayed in Fig. 4.2 and Table 4.1. When all these corrections have been applied, the radiocarbon dates are known as 'calibrated ages', and are quoted as calendar dates (BC and AD).

4.3 Western Crete Data

In earlier studies of the AD 365 uplifted shoreline, marine organisms (mostly encrusting algae and vermetids, but not lithophagids) were collected at a number of locations close to the top of the paleo-shoreline and were radiocarbon-dated, largely by conventional (non-AMS) methods [7, 8, 29–32]. These measurements, which were recalibrated by Price et al. [33] in consistency with the dates reported here, are also displayed in Fig. 4.2. The youngest calibrated ages, obtained from the encrusting algae [7, 8, 29–32], lie within ~200 years of AD 365 consistent with the suggestion that at least the upper part of the exposed marine substrate below the paleo-shoreline was lifted up during that event.

Additionally, I found that 11 out of 13 new radiocarbon ages for corals and bryozoans bracket AD 365 within their 2σ error bounds, and because the samples covered almost all of the range between present sea-level and the paleo-shoreline (Fig. 4.2, Table 4.1), I concluded in Chap. 2 that *all* of the observed uplift occurred in a single event in AD 365.

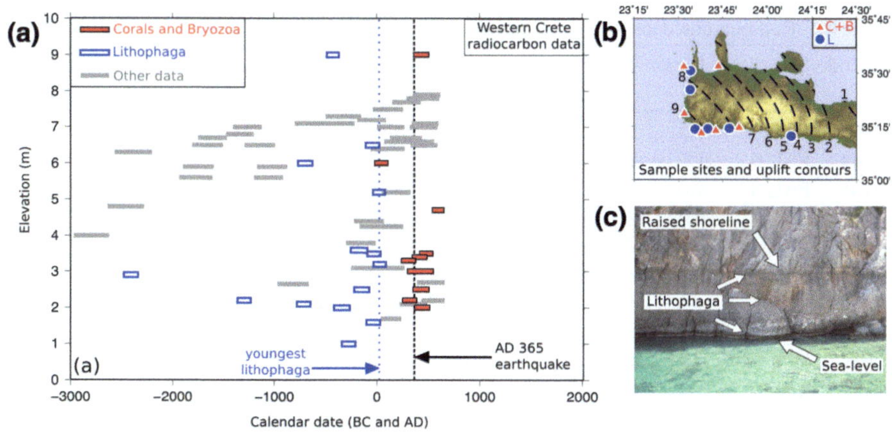

Fig. 4.2 **a** Radiocarbon dates from Western Crete. *Grey bars* are radiocarbon dates obtained prior to the Shaw et al. [34] study described in Chap. 2; see text for references. The youngest ages coincide with AD 365. *Red bars* are coral and bryozoan ages from Shaw et al. [34], nearly all closely bracketing AD 365. Lithophagid dates (*blue bars*) are offset from the coral and bryozoan ages, and have a much larger range. All *bars* are 1σ errors. **b** Contours of uplift, in metres, of the AD 365 paleo-shoreline (Figs. 2.3, 2.4), with sample sites for corals and bryozoans (*red*) and lithophagids (*blue*). **c** The AD 365 paleo-shoreline on the south coast, marked by a white band of algal encrustation. The same white band is forming today at sea-level. Between the top of the shoreline and present sea-level, the cliff face is covered with lithophagid borings

Confirmation of the date of death of the corals comes from U/Th dating (carried out by A. Thomas and G. Henderson at the University of Oxford). Three of the coral samples were split into two pieces; one half was sent for radiocarbon dating, and the other half was dated using U/Th methods. In spite of a high initial Th concentration (probably due to incorporation of siliceous mud in the carbonate skeleton) and consequently relatively large error bounds, the U/Th ages were consistent with the radiocarbon ages, confirming that the radiocarbon ages obtained from the corals and bryozoans accurately record the date of death of the organisms (Fig. 4.3; Tables 4.2, 4.3).

I also collected 14 samples of *Lithophaga lithophaga* which had been preserved in their boreholes between the top of the uplifted shoreline and present-day sea-level (blue bars in Fig. 4.2a, blue dots in Fig. 4.2b, Table 4.1). *Lithophaga lithophaga* [37] is a species of marine bivalve belonging to the mussel family (Mytilidae), which bores into calcareous substrates using acid secreted from its pallial gland [38, 39]. It can live between the sea surface and around 30 m depth, although the highest concentrations are found within the top few metres, and their upper limit coincides closely with sea-level [40]. Figure 4.2c is a field photograph which shows the typical appearance of the uplifted paleo-shoreline (see also Chap. 2, Figs. 2.3, 2.4). The exposed cliff-face below the raised shoreline is densely covered by the borings of *L. lithophaga* and, in some holes, shells can be found *in situ*.

Since the lithophagids collected in western Crete grew on the same uplifted marine substrate as the corals and bryozoans whose ages closely concentrate near AD 365,

Table 4.1 Radiocarbon dates of Cretan samples

Labcode	Site	Height m	Material	^{14}C age BP	±	Calibrated range (68.2% prob.)	Calibrated range (98.4% prob.)
16743	PHA 1	3.5	Coral (S)	1985	28	AD 395–AD 522	AD 327–AD 576
16744	PAL 6	1.6	*Lithophaga*	2396	30	BC 101 – AD 32	BC 165–AD 84
16745	PAL 8	9.0	Coral (B)	2019	28	AD 338–AD 468	AD 275–AD 538
16788	PAL 3	9.0	*Lithophaga*	2715	28	BC 486–BC 366	BC 601–BC 325
16991	PHA 3	3.0	Coral (S)	2032	28	AD 326–AD 451	AD 264–AD 521
16993	KIR 1	2.2	*Lithophaga*	3401	31	BC 1366–BC 1232	BC 1407–BC 1146
17017	SFI 1	2.0	*Lithophaga*	2660	27	BC 416–BC 261	BC 486–BC 210
17018	PHA 2	3.5	*Lithophaga*	2390	27	BC 92–AD 37	BC 157–85
17019	KOU 1	2.5	*Lithophaga*	2490	28	BC 219–BC 72	BC 322–BC 31
17020	SOU 1	2.9	*Lithophaga*	4293	30	BC 2480–BC 2339	BC 2556–BC 2278
17269	PAL 1	5.2	*Lithophaga*	2350	29	BC 37–AD 81	BC 111–AD137
17270	SOU 1	6.5	*Lithophaga*	2403	28	BC 109–AD 24	BC 168–AD 75
17271	PAL 5	3.6	*Lithophaga*	2506	29	BC 252–BC 93	BC 332–BC 47
17272	PAL 7	6.0	*Lithophaga*	2904	29	BC 768–BC 626	BC 791–BC 527
17273	HSF 2	2.1	*Lithophaga*	2914	31	BC 780–BC 646	BC 797–BC 535
17330	KIR 2	3.2	*Lithophaga*	2339	29	BC 31–AD 90	BC 101–AD 146
17331	HSF 1	1.0	*Lithophaga*	2578	35	BC 339–BC 206	BC 381–BC 139
17669	CHS 1	3.3	Bryozoans (My)	2129	27	AD 215–AD 350	AD 151 – AD 401
17670	CHS 2	3.0	Bryozoans (My)	2071	27	AD 280–AD 405	AD 222–AD 460
17671	CHS 5	2.5	Bryozoans (My)	2019	27	AD 340–AD 465	AD 277–AD 536
17672	CHS 5	2.5	Bryozoans (My)	2024	26	AD 335–AD 459	AD 272–AD 530
17673	CHS 7	6.0	Coral (S)	2345	27	BC 33–AD 83	BC 104–AD 140
17674	AGM	3.0	Coral (S)	1977	26	AD 409–AD 531	AD 340–AD 580
17675	SOU NB2	2.2	Bryozoans (My)	2119	29	AD 230–AD 360	AD160–AD 414
17676	GRAM 2	4.7	Bryozoans (My)	1878	25	AD 514–AD 630	AD 452–AD 658
17677	GRAM 4	2.0	Bryozoans (My)	2015	28	AD 342–AD 472	AD 281–AD 541
17678	PHA 2	3.4	Coral (S)	2028	27	AD 331–AD 455	AD 268–AD 525

S *Stenocyathus vermiformus*, B *Balanophyllia regia*, Ca *Caryophyllida*, My *Myriapora truncata*. Sample sites are listed in Chap. 9.

their ages might also be expected to bracket the same date. All of the lithophagid ages are older than AD 365, and are therefore consistent with uplift in the earthquake, but they have a much larger range (2,107 years) than the range (640 years) of coral–bryozoan ages and, in addition, the average coral–bryozoan ages are also ~350 years younger than the youngest lithophagid ages, suggesting a possible systematic offset in the lithophagid ages (Fig. 4.1a). In the following sections, I investigate two possible reasons for the observed difference in coral–bryozoan and lithophagid ages.

4.4 Colonisation Order and Preservation Potential

Lithophagids colonise calcareous substrates and excavate their own boreholes, beginning at the surface as juveniles and boring deeper into the limestone as adults. The entrances to the boreholes are therefore commonly narrower than the adult shell,

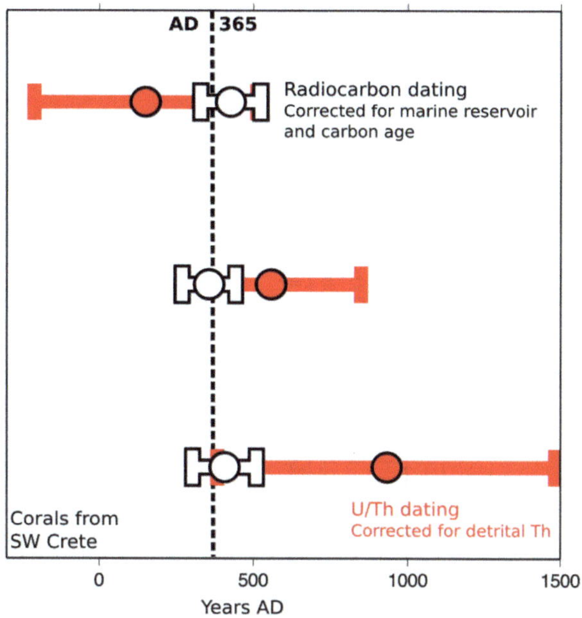

Fig. 4.3 Radiocarbon (*white*) and U/Th (*red*) dating of corals from Western Crete. *Error bars* are 1σ

which means that the shells are often difficult to dislodge after death, thereby greatly enhancing their preservation potential. Cross-cutting relationships between generations of lithophagids can be recognised in Kynos, on the southern coast of the Gulf of Evia, where later generations have bored into cement-infilled earlier holes (Fig. 4.4a, b) showing that the same area of cliff-face can be colonised repeatedly over an extended period. In an example from western Crete, the shell of a lithophagid provided protection for another elongated bivalve species, which was found occupying the vacated shell (Fig. 4.4c, d). Many of the corals and bryozoans that were collected in western Crete were also found within the protected vacated boreholes of lithophagids (Fig. 4.4e), showing that there were generations of lithophagids present in the cliffs before the corals and bryozoans settled there.

By contrast, in every place that uplifted corals and bryozoans were observed, they were clearly late colonisers, belonging to the very last phase prior to uplift. They grew on top of all other encrusters and borers, and have a relatively low preservation potential because of their delicate, stick-like structure and because they attach to and protrude from the limestone substrate, making them much easier to dislodge than the lithophagid shells (Fig. 4.4e).

If the difference in colonisation time and preservation potential were the sole cause of the discrepancy in radiocarbon ages between the corals-bryozoans and the lithophagids, we might expect to see the date of uplift well constrained by the corals and bryozoans, whereas the lithophagids should have a wide range of ages, from about 6,000 years BP (when sea-level reached relative stability near its present level following the last glacial maximum [42–44]) to the date of uplift. The lithophagids do

Table 4.2 Uranium and thorium isotope data

Sample	Mass (g)	^{238}U (ppm)	±	^{232}Th (ppb)	±	($^{230}Th/^{238}U$)	±	($^{234}U/^{238}U$)	±	($^{232}Th/^{238}U$)	±
AGM	0.0874	4.480	0.001	103.9	0.2	0.027	0.002	1.1450	0.0008	0.00759	0.00002
PAL 08	0.2221	3.692	0.003	75.9	0.3	0.022	0.001	1.1444	0.0008	0.00673	0.00002
PHA 01	0.0312	3.868	0.001	106.1	0.6	0.020	0.004	1.1401	0.0008	0.00898	0.00005

Chemical purification of U and Th and measurement by MC-ICP-MS was by standard protocols [35]. Ratios are presented as activity using the half lives of Cheng et al. [36]

Table 4.3 Conventional and calibrated radiocarbon ages, for details of calibration see text

Sample	C-14 age year BP	±	Corrected radiocarbon date	U/Th age uncorrected	±	U/Th age corrected	+	−	U/Th corrected date	+	−
AGM	1977	26	AD 340–AD 580	2530	180	1840	320	410	AD 106	317	405
PAL 08	2019	28	AD 275–AD 538	2040	90	1440	250	330	AD 514	247	333
PHA 01	1985	28	AD 327–AD 576	1890	410	1060	510	590	AD 891	891	591

U/Th age uncorrected is calculated using data from Table 4.2 and the Isoplot software [41]. Corrected ages account for the presence of initial Th, assuming the initial Th has an isotopic composition of continental material. The [$^{232}Th/^{230}Th$] of the initial Th component is taken from the GEOROC database http://georoc.mpch-mainz.gwdg.de/georoc/ (median [$^{232}Th/^{230}Th$] = 179,621) uncertainty in this value is estimated from the 5th and 95th percentile of the [$^{232}Th/^{230}Th$] from the database ([$^{232}Th/^{230}Th$]0.05 = 121,411 and [$^{232}Th/^{230}Th$]0.95 = 273,563). All ages are presented as years before AD 1950

Fig. 4.4 Repeated colonisation of lithophagid borings. **a**, **b** Cross-cutting between lithophagid boreholes, with young shells boring into the infill of older holes, confirming that multiple generations of lithophagids lived on the cliff face prior to uplift (photographed at Kynos; see Fig. 4.7 for location). **c**, **d** An elongated bivalve occupying the shell of an older lithophagid (Paleochora). **e** Representative photograph of the habitat of the corals and bryozoans: in all cases they were the most recent colonisers, and in this case were found within a lithophagid borehole. **f** Lithophagid holes bored into recent algal encrustation in western Crete

indeed display a much wider range of ages than the corals and bryozoans, but there is also an offset between the youngest lithophagid ages and the average coral–bryozoan ages of around 350 years. Colonisation sequence and preservation can explain the larger range of the lithophagid ages compared with coral–bryozoan ages, but cannot explain the systematic offset.

4.5 Incorporation of Old Carbonate into Shells

In general, there are two possible sources for the carbon that bivalves use to build their shells: metabolic carbon, which comes from food, and dissolved inorganic carbon, which comes from the surrounding sea water. Shell-building has been extensively studied in the edible mussel *Mytilus edulis*, which builds its shells from less than 10% metabolic carbon [45]. *Mytilus edulis* is in the same family as *Lithophaga* and this value seems to be a reasonable average for bivalves generally, although the maximum observed proportion of metabolic carbon in a bivalve shell is 35% [46]. The remainder of the shell carbon, and therefore the majority, comes from HCO_3^- and CO_3^{2-} in the water column.

A complexity arises in the case of lithophagids, because they bore chemically into their substrate, releasing bicarbonate ions whose ^{14}C activity reflects the age of the host rock, not the organism. If any of this carbon is incorporated into the lithophagid shell, then the proportion of ^{14}C to ^{12}C in the shell will be lower than a shell that had grown from seawater alone, resulting in an artificially old radiocarbon age. Thus, if the limestone substrate acts as a carbon source for their shells, those shells will have a radiocarbon age that is older than their true age.

This effect is analogous to the 'hard water effect', which has been recognised for some time (e.g. [47, 48]) and affects the radiocarbon dating of lacustrine molluscs. That effect arises when ^{14}C is diluted by dissolved bicarbonate from ancient limestone rocks which make up the catchment area of the lake, and again results in an artificially old age when the mollusc shells are radiocarbon-dated [49].

In the next section, I investigate this effect using modern lithophagid samples.

4.5.1 Test Using Modern Lithophagids

The simplest test of whether lithophagids have artificially old ages due to the incorporation of old, detrital carbon, would be to radiocarbon-date the shell of a live lithophagid collected from present sea-level and compare its age directly with that expected from modern atmospheric ^{14}C levels. This is complicated, however, by enriched levels of ^{14}C from nuclear testing that have entered the oceans, but at variable concentrations that depend on ocean depth and latitude [50].

I therefore carried out a test using lithophagids that were collected alive, with a well-recorded collection date and location, prior to 1950. I located four museum specimens which were thought to have been collected alive from tectonically stable or subsiding areas (Table 4.4; Fig. 4.5). Three of these samples were *Lithophaga lithophaga* shells from the central and western Mediterranean. The fourth sample was a *Lithophaga obesa* from the Red Sea, which had been entirely preserved in ethanol, including its soft tissues. All were analysed at the Scottish Universities Environment Research Centre (SUERC) at East Kilbride, and radiocarbon ages were corrected to obtain calibrated ages in the exactly the same way as for the Cretan samples. All four samples were extremely clean, articulated, with a complete periostracum and nacreous lustre inside the shell, suggesting that they were either alive or very recently dead when collected.

The calibrated radiocarbon ages of the museum samples are plotted relative to their date of collection (and, one assumes, their date of death) in Fig. 4.5d. All of the modern *L. lithophaga* shells have ages that are offset from the date of death by approximately 900–1,400 years. The date obtained from the *L. obesa* shell is also older than the date of death but, in contrast with the other modern samples, is offset by just 200 years.

The calibrated ages of the lithophagids I collected in western Crete are also plotted at the bottom of Fig. 4.5d (on the horizontal axis) as open triangles relative to AD 365 (that is, AD 365 would plot at the zero or 'date of death'). The range of offsets from AD 365 (approximately 300–2,800 years) obtained from these samples is shown as a grey region. All of the museum lithophagid samples plot within this grey field. The Cretan coral and bryozoan data are plotted at the top of this figure (black triangles), and plot close to zero, showing that the coral and bryozoan ages are not significantly offset from AD 365.

Table 4.4 CZM: Cambridge Zoology Museum, NHM: Natural History Museum, London

Labcode	Collection site	Collection date	Species	^{14}C age year BP	Calibrated date 68.2%	Calibrated date 95.4%	Source
SUERC-18952	Egypt	1934	*L. obesa*	638 ± 35	AD 1653–1805	AD1567–1860	CZM
SUERC-18953	Bay of Naples	1918	*L. lithophaga*	1875 ± 37	AD 511–638	AD 445–665	NHM
SUERC-18954	Sardinia	1905	*L. lithophaga*	1783 ± 37	AD 604–697	AD 550–764	NHM
SUERC-18955	Trieste	1890s	*L. lithophaga*	1515 ± 37	AD 864–994	AD 793–1030	NHM

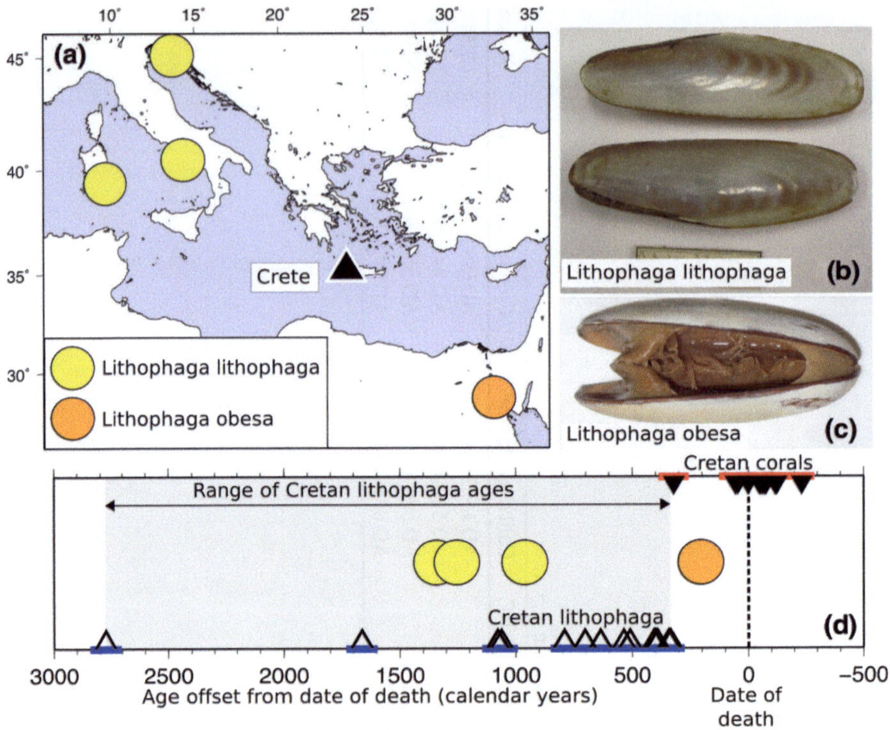

Fig. 4.5 **a** Locations of *Lithophaga lithophaga* (*yellow*) and *Lithophaga obesa* (*orange*) museum samples collected from the Mediterranean and Red seas. **b, c** Photographs of one of the *L. lithophaga*, and the *L. obesa* samples, both are ∼5 cm long. **d** Calibrated ages relative to the date of collection. If the museum lithophagid samples recorded the correct age, they would plot at zero. The *L. lithophaga* (*yellow*) are offset 1,000–1,400 years from the date of collection, whereas the *L. obesa* is offset only 200 years. The calibrated radiocarbon ages of the Cretan lithophagids from Chap. 2 are plotted as *white triangles* along the *bottom* of the figure relative to AD 365: if they correctly recorded the date of their uplift, they should plot at zero. The age offset of the museum lithophagid specimens plots within the range of lithophagid ages from Crete. The calibrated radiocarbon ages of the corals and bryozoans from Chap. 2 are plotted as *black triangles* at the *top* of the figure, again relative to AD 365. They plot at the origin, and therefore correctly record their date of uplift and death

The modern samples confirm that lithophagids can record an artificially old age when radiocarbon-dated. Intriguingly, the *L. obesa* age is offset by a much smaller amount than any of the *L. lithophaga* dates, which may be related to differences in their ecology. *L. obesa* preferentially bores into dead coral heads [51, 52], which are likely to be very young, whereas *L. lithophaga* bores into a variety of carbonate substrates, from Mesozoic limestone (Fig. 4.4a, b) to modern algal encrustations (Fig. 4.4f). The difference in age offset may therefore depend on the age of the substrate into which the lithophagid is boring, and the consequent depletion of that substrate's ^{14}C. In the following section, I apply these insights to the lithophagid ages from western Crete.

4.5.2 Estimating the Amount of Old Carbon Incorporated

Consider a hypothetical situation in which a coral and a lithophagid, which were both killed in an uplift event T years ago, are dated using radiocarbon methods.

For the radiocarbon dating, the modern-day fraction, R, of $^{14}\mathrm{C}/^{12}\mathrm{C}$ is measured, i.e. R_C is the measured ratio in the coral and R_L is the ratio in the lithophagid. According to the decay equation, R_C and R_L follow the relation

$$R = A\exp\left(-\frac{T}{8033}\right) \tag{4.1}$$

where A is the initial $^{14}\mathrm{C}/^{12}\mathrm{C}$ ratio of the medium from which the organisms grew, and T is the time since death.

If one assumes that the coral and lithophagid both acquired their carbon solely from seawater, and therefore have an initial $^{14}\mathrm{C}/^{12}\mathrm{C}$ equal to that of seawater, then the measured R_C and R_L will produce an apparent difference in date of death between the coral and lithophagid of Δ years. The difference between the measured R_C and R_L values arises because, in reality, the lithophagid sourced its carbon both from seawater and from 'dead' (i.e. ^{14}C-free) carbon from the limestone into which it had bored. The lithophagid therefore had a different initial $^{14}\mathrm{C}/^{12}\mathrm{C}$ from the coral.

One can determine the initial $^{14}\mathrm{C}/^{12}\mathrm{C}$ ratio of the lithophaga, A_L as follows:

$$R_L = A\,e^{-(T+\Delta)/8033} = A\,e^{-\Delta/8033}e^{-T/8033} = A_L\,e^{-T/8033} \tag{4.2}$$

so

$$A_L = A\,e^{-\Delta/8033} \tag{4.3}$$

The motivation of this calculation is to determine the fraction, γ of shell carbon that is sourced from the limestone substrate into which the lithophagid has bored, given an age offset of Δ. The composition of carbon in seawater is

$$C_W = \alpha\,^{12}\mathrm{C} + (1-\alpha)\,^{14}\mathrm{C} \tag{4.4}$$

$$= \alpha\left(^{12}\mathrm{C} + A\,^{14}\mathrm{C}\right) \tag{4.5}$$

where α is the proportion of ^{12}C. Here, the proportion of ^{13}C is ignored, since it is irrelevant to the calculation. The limestone substrate consists entirely of ^{12}C, so the carbon composition of the rock is

$$C_R = {}^{12}\mathrm{C} = \alpha\,(1+A)\,^{12}\mathrm{C} \tag{4.6}$$

since $\alpha\,(1+A) \equiv 1$.

The composition of carbon within a living lithophagid is given by

$$C_L = \gamma C_R + (1-\gamma)\, C_W \tag{4.7}$$

$$= \alpha \left[\gamma\,(1+A)\,{}^{12}\mathrm{C} + (1-\gamma)\left({}^{12}\mathrm{C} + A^{14}\mathrm{C}\right)\right] \tag{4.8}$$

$$= \alpha \left[(1+A\gamma)\,{}^{12}\mathrm{C} + (1-\gamma)\,A^{14}\mathrm{C}\right]. \tag{4.9}$$

Therefore the initial ${}^{14}\mathrm{C}/{}^{12}\mathrm{C}$ within the lithophaga is given by

$$A_L = \frac{1-\gamma}{1+A\gamma} A \tag{4.10}$$

and

$$e^{-\Delta/8033} = \frac{1-\gamma}{1+A\gamma}. \tag{4.11}$$

The ratio A is $\sim 10^{-12}$ and γ is a small fraction, therefore the product $A\gamma \ll 1$ can be ignored. The proportion of carbon within the lithophagid that originates from limestone is given by

$$\gamma \approx 1 - e^{-\Delta/8033}. \tag{4.12}$$

If $\Delta = 500$ years, approximately the offset seen between the average age of the coral–bryozoan samples and the youngest lithophagid samples from Crete (see Fig. 5.1), then $\gamma = 6\%$.This is a small fraction, and seems plausible for the amount of 'dead' carbon incorporated in the shell. Even if $\Delta = 1{,}000$ years, $\gamma = 12\%$. Therefore only a small fraction of shell carbon must be sourced from the limestone substrate in order for the age offset of the lithophagid to be significant. Figure 4.6 is a diagrammatic summary of the various sources of shell carbon for a lithophagid, based on the calculation above.

In fact, with one exception (sample 17020, Table 4.1) the age offsets for the lithophagids relative to the corals and bryozoans from Crete are smaller than ~1,600 years, and all the modern lithophagids have age offsets smaller than 1,400 years. These differences are small compared with the decay constant for ${}^{14}\mathrm{C}$, so one can make use of the linear approximation that each 1% of 'dead' carbon incorporated into the sample is equivalent to an apparent increase in the samples age of 1% of the decay constant, or 80 years. Thus three of the modern lithophagid samples and 4 out of the 14 samples from western Crete, show age discrepancies between about 800 and 1,600 years, equivalent to an incorporation of 10–20% of dead carbon from the host rock.

However, a substantial subset of 7 of the 14 lithophagids from western Crete show smaller age offsets between 530 and 330 years. This could imply smaller contributions (4–7%) of dead carbon to their shells. But, alternatively, some of these samples may have incorporated carbon from a Holocene algal marine encrustation that commonly overlies the Mesozoic carbonate (as illustrated in Fig. 4.4f). This encrustation could not have formed before the sea surface stabilised near its present level at about 6 ky BP [42–44], so its maximum age at the time of the AD 365 earthquake would have been about 4,500 years and, assuming a constant rate of encrustation, its average age would have been about 2,250 years. The average ^{14}C activity of this encrustation would therefore have been approximately 75% of the activity of the open ocean, so the influence on the apparent age of the lithophagids of incorporating 1% of carbon of this activity would be approximately one-quarter the influence of incorporating the same proportion of dead carbon. If, therefore, the lower range of lithophagid age offsets arises from the incorporation of carbonate from the Holocene encrustation, it would require a greater proportion (16–30%) of the lithophagid shell to be derived from the host rock. Thus contamination from younger encrustation as well as from Mesozoic limestone allows yet further spreading of lithophagid ages. More detailed analysis of the present dataset is not justified: most of the samples collected had probably bored through at least some encrustation into the underlying old limestone, but in some cases the encrustation had fallen off the cliff-face.

I have no objective way of distinguishing age offsets induced by the incorporation of dead carbon from those produced by the preservation of lithophagid shells *in situ* long after death. Indeed, both effects are likely to occur simultaneously. The existence of a lower bound (~400 years) on the age offsets observed in Crete requires some dilution of environmental carbon by material from the host rock. The minimum proportion of host-rock material required to explain that offset is about 5% if all that material is carbon-dead. The age offsets on modern lithophagids require between 10 and 20% of dead carbon, which is comparable to the fraction estimated from the smallest lithophagid age offsets from western Crete if they were generated entirely by the incorporation of carbon from the Holocene encrustation.

Regardless of the details of their explanation, it must be accepted that there are offsets of up to 2,000 years in the ages of lithophagids. I now consider the implications of these findings for tectonics and paleoseismology.

4.6 Implications for Tectonics and Paleoseismology

In western Crete, the large amount of uplift in the single big earthquake of AD 365, confirmed by both radiocarbon dates and the physical appearance of the uplifted paleo-shoreline, has allowed me to identify two effects that result in artificially old estimates of the date of that uplift event from lithophagid radiocarbon ages. The first is a combination of colonisation order and the high potential for lithophagid shells to be preserved *in situ* in their bore-holes long after their death. The second effect is the incorporation into the shell of ^{14}C-free carbon from the limestone substrate. The age offset from this effect alone is at least 350 years on Crete, and at least 800 years for the museum specimens. In this section I illustrate the implications of these effects in a different tectonic environment, in the extensional graben system of central Greece.

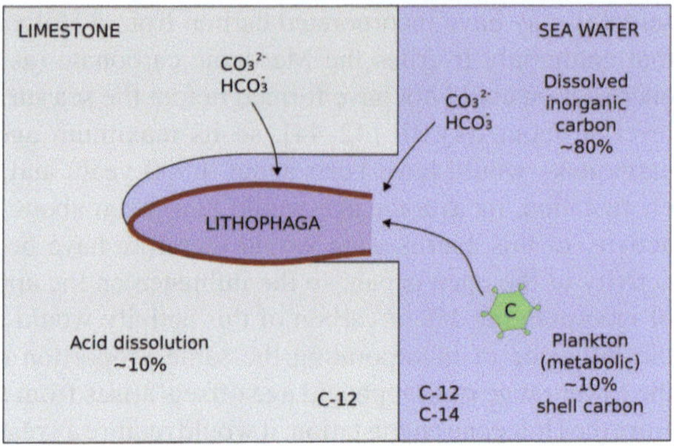

Fig. 4.6 Potential carbon sources for the shell of *L. lithophaga*. Most of its shell carbon is from a combination of dissolved inorganic carbonate in sea water and metabolic carbon from food. The Cretan samples suggest that ~10% of the carbon comes from the host limestone into which the shell is bored (see text). The proportions from sea water and food are based on those known from *Mytilus edulis*

The Gulfs of Corinth and Evia in Central Greece (Fig. 4.6) are large-scale active normal-faulting systems (e.g. [53]). Uplift occurs in the footwalls of these faults during earthquakes with repeat times of roughly 200–300 years in Corinth [54, 55] and 1,000–2,000 years in Evia [56]. Extension rates across the graben measured by GPS are consistent with these paleoseismological estimates of repeat times [57, 58]. With maximum fault-segment lengths of about 20 km, the earthquakes are much smaller than the AD 365 Crete event, and are typically up to M_w 6.0–6.5, with perhaps just 10 cm uplift in each event [5]. The Late Quaternary rate of uplift in the Gulf of Corinth can be estimated from matching flights of uplifted marine terraces to sea-level stillstands, and from U/Th dating of corals found on those terraces (e.g. [6, 59–61]).

A series of marine notches can also be observed in the footwalls of the recently active normal faults, particularly in the Gulf of Corinth (Fig. 4.6, photographs). The elevation of the highest notch varies from east to west, with the highest elevations (~10 m) associated with the fastest extension rates (~12 mm year^{-1}) in the west; both are lower in the east (~3 m and $<$ 5 mm year^{-1} respectively; e.g. [58, 62, 63]). The cliff faces beneath the notches are punctured by lithophagid borings, some of which have been collected and radiocarbon-dated to investigate uplift rates over the last few thousand years [13, 64–71].

Lithophagid holes and associated notches are also seen in the Gulf of Evia, to a lower maximum height of typically ~1 m. Here, older lithophagid ages beneath higher younger ones have been used to infer subsidence followed by uplift at the same site over a few hundred years [65, 70]). In the light of the insights from western Crete, it is not surprising that we see reversals in radiocarbon ages of lithophagids with elevation caused by prolonged and repeated colonisation of an exposed substrate, and the examples in Fig. 4.4a and b were taken from this site. With lithophagid shells remaining intact and *in situ* for a thousand or more years after death, a complicated history of reversed vertical motions seems poorly supported by such evidence.

Goldsworthy and Jackson [63] suggested that the highest notches and the highest exposed lithophagid boreholes in the Gulfs of Corinth and Evia were formed at about

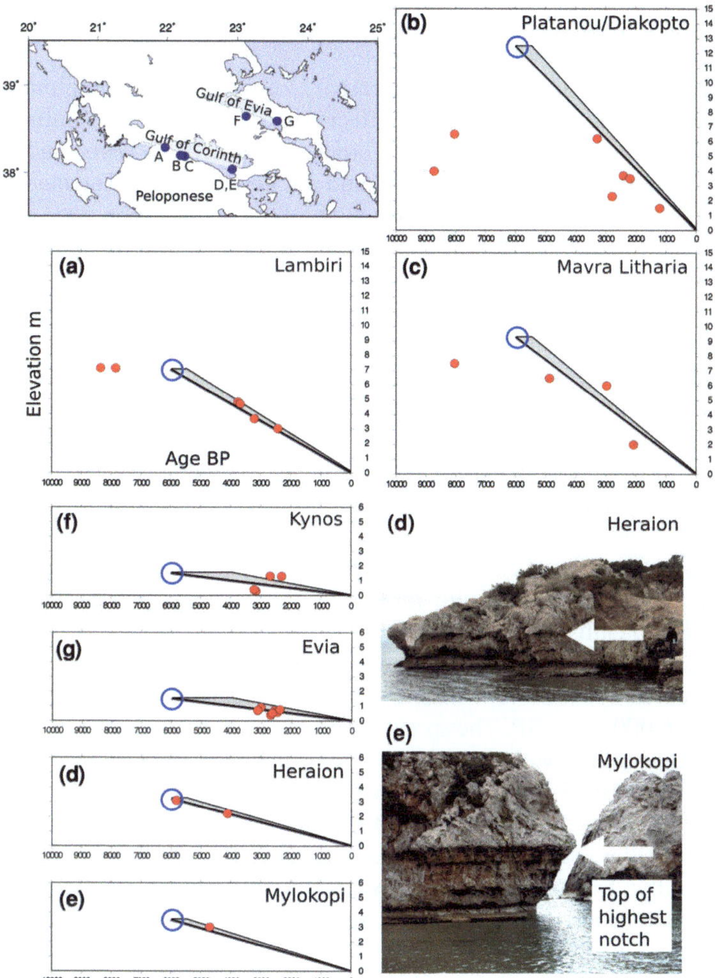

Fig. 4.7 Conventional radiocarbon ages of *L. lithophaga* (*red dots*) collected from the Gulfs of Corinth and Evia plotted against elevation above present sea-level. The location of each plot, and of the field photos, is identified by *letter* on the map. *Blue circles* mark the elevation of the top of the highest marine notch or boring in each locality. At Lambiri, the top of the cliff sequence is buried beneath road talus, and is only known to be higher than 7 m. If the highest notch formed 6000 years BP, then the *black line* represents the average uplift rate over that period. The *grey triangle* is the range of average uplift rates allowing for local earthquake repeat times (see text). Note that the lithophagid ages plot on, or to the left of, these lines. Field photos show multiple notches at Heraion and Mylokopi, with *white arrows* marking the position of the highest one

6,000 years BP, when sea-level stabilised near its present level, having risen at very rapid rates of up to 1–2 m per century over the previous 10,000 years. If correct, this rule-of-thumb provides a very useful quick estimator of uplift in the late Holocene. In

Fig. 4.7 I examine the extent to which this suggestion is compatible with published lithophagid ages.

In Fig. 4.7 conventional (i.e. uncalibrated) radiocarbon ages are plotted because the appropriate correction for reservoir effect is poorly known in the Gulfs of Corinth and Evia, but since this effect is of the order of hundreds of years, and the x-axis in Fig. 4.7 is in thousands of years, it should not be significant. The map shows the sample locations, identified as A–G on the plots of radiocarbon ages and in field photos. In the plots of radiocarbon age against elevation, the lithophagid ages are shown as red dots. The blue circles mark the elevation of the top of the highest notch or limit of lithophagid holes in each locality. At Lambiri, the top of the cliff sequence is buried beneath road talus, and is only known to be higher than 7 m. If we assume in each place that the highest notch or lithophagid holes formed at 6,000 years BP, the black line joining the blue circle to the origin represents the average, smoothed, long-term uplift rate. The lithophagid ages should plot to the left of this line. Some uncertainty arises because the first earthquake at a site may not have occurred immediately when sea-level stabilised, but will presumably have occurred somewhere between 6,000 years BP and a following time interval corresponding to the average repeat time of earthquakes at that location. The grey triangles therefore bound the expected average uplift rates, based on the expected earthquake repeat times in the two Gulfs.

The majority of uncalibrated lithophagid ages plot either on top, or to the left, of the estimated uplift curve, consistent with the suggestion that the highest notch formed at ~6,000 years BP. The spread of lithophagid ages relative to the expected age at that elevation (given by the black line) is large in places, with the oldest ages up to 9,000 years BP (Fig. 4.7a–c). Some of this variation is expected if lithophagids are preserved *in situ* for centuries to millennia after death while the substrate is colonised by other organisms; but the oldest ages of 8,000–9,000 years cannot be explained in this way. At 8,000–9,000 years BP, before it reached relative stability near its present level ~6,000 years ago, sea-level was up to 25 m lower than today. At that level it would have been impossible for lithophagids to live at the position they now occupy on the substrate without uplift rates a factor of 3 or more faster than the Late Quaternary averages based on the correlation of terrace heights with marine stillstands, which would also be much faster than expected from fault slip-rates that are consistent with GPS and paleoseismology. These older ages, roughly 2,000–3,000 years beyond the expected 6,000 years BP maximum, suggest that the lithophagid age offset in the Gulf of Corinth due to the incorporation of ancient limestone substrate carbon into shells may be as much as 2,000–3,000 years.

4.7 Conclusions

The ubiquity of lithophagids in the Mediterranean, especially the species *Lithophaga lithophaga*, together with their high potential for preservation, makes them very useful indicators of uplift in tectonically active areas. Quantifying that uplift requires some estimate of its age, and dates determined from radiocarbon ages on lithophagids have often been used for this purpose. The occurrence of a single, large-amplitude uplift event of up to 10 m in western Crete described in Chap. 2, associated with a very large ($M_w \geq 8$) earthquake in AD 365, reveals two effects that limit the resolution and applicability of radiocarbon lithophagid ages for tectonic and paleoseismological purposes.

The first effect is that the good preservation potential of lithophagids can cause them to be preserved *in situ* long after death, while the marine substrate continues to be colonised by other marine organisms, including other lithophagids. The second effect is a systematic error in the lithophagid age caused by the incorporation of 'dead' (i.e. ^{14} C-free) carbon into the shell from the host limestone into which it bored, which can lead to ages up to 1,000–2,000 years too old, even in museum specimens collected before the atmospheric testing of nuclear weapons. In western Crete the second effect is probably responsible for a minimum offset of about 350 years between the youngest lithophagid ages and the ages of corals and bryozoans that faithfully record the AD 365 uplift and were the last colonisers of the substrate.

These two effects mean that, even where the uplift occurs in a single massive event, as in AD 365 in Crete, lithophagid ages alone are unlikely to constrain the age of the uplift to better than a thousand years, or even to demonstrate that it all occurred at once. In places like the Gulf of Corinth, where uplift occurs in events of much smaller magnitude ($\sim$10 cm vs. $\sim$10 m) separated by hundreds of years, lithophagid ages are clearly unable to distinguish individual uplift events for paleoseismological purposes. They can, however, be used to test (but not prove) the hypothesis that the highest uplifted lithophagid borings and marine notches in the coastal regions of central Greece were formed about 6,000 years BP, when sea-level became relatively stable near its present level, following a rapid rise at the end of the last glacial maximum. The available lithophagid ages in Greece are indeed consistent with that suggestion.

References

1. R. Wallace, *Active Tectonics, Studies in Geophysics* (National Academy Press, Washington DC, 1986), p. 265
2. R. Yeats, K. Sieh, C. Allen, *The Geology of Earthquakes* (Cambridge University Press, Cambridge, 1996), p. 568.
3. D. McKenzie, Active tectonics of the Alpine–Himalayan belt: The Aegean sea and surrounding regions. Geophys. J. Int. **55**, 217–254 (1978)
4. S. McClusky, Global positioning system constraints on plate kinematics and dynamics in the eastern Mediterranean and Caucasus. J. Geophys. Res. **105**, 5695–5719 (2000)
5. J. Jackson, J. Gagnepain, G. Houseman, G. King, P. Papadimitriou, C. Soufleris, J. Virieux, Seismicity, normal faulting, and the geomorphological development of the Gulf of corinth (Greece): The Corinth eathquakes of February and March 1981. Earth Planet. Sci. Lett. **57**, 377–397 (1982)
6. R. Armijo, B. Meyer, G. King, A. Rigo, D. Papanastassiou, Quaternary evolution of the Corinth rift and its implications for the late Cenozoic evolution of the Aegean. Geophys. J. Int. **148**, 596–618 (1996)

7. J.A.P. Laborel, J. Thommeret, Y. Thommeret, Holocene raised shorelines in western Crete (Greece), Proceedings 1978 International Symposium on Coastal Evolution in the Quaternary (Sao Paulo, 1979), pp. 475–501
8. Y. Thommeret, J. Thommeret, J. Laborel, L.F. Montaggioni, P.A. Pirazzoli, Late Holocene shoreline changes and seismo-tectonic displacements in western Crete (Greece). Zeitschrift für Geomorphologie Suppl. **40**, 127–149 (1981)
9. P.A Pirazzoli, J. Laborel, S.C. Stiros, Earthquake clustering in the eastern Mediterranean during historical times. J. Geophys. Res. **101**, 6083–6097 (1996)
10. P.M. De Martini, D. Pantosti, N. Palyvos, F. Lemeille, L. McNeill, R. Collier, Slip rates of the Aigion and Eliki faults from uplifted marine terraces, Corinth Gulf, Greece. Comptes Rendus Geosci. **336**, 325–334 (2004)
11. L.C. McNeill, R.E.L.I. Collier, Uplift and slip rates of the eastern Eliki fault segment, Gulf of Corinth, Greece, inferred from Holocene and Pleistocene terraces. J. Geol. Soc. Lond. **161**, 81–92 (2004)
12. F. Laborel, J. Laborel-Deguen, Biological indicators of relative sea-level variations and of coseismic displacement in the Mediterranean region. J. Coast. Res. **10**, 395–415 (1994)
13. S. Papageorgiou, M. Arnold, J. Laborel, S. Stiros, Seismic uplift of the harbour of ancient Aegira, Central Greece. Int. J. Naut. Archaeol. **22**(3), 275–281 (1993)
14. A.I. Galinou-Mitsoudi, S. Sinis, Age and growth of Lithophaga lithophaga (Linnaeus, 1758) (Bivalvia: Mytilidae), based on annual growth lines in the shell. J. Molluscan Stud. **61**, 435–453 (1995)
15. W.F. Libby, E.C. Anderson, J.R. Arnold, Age determination by radiocarbon content: World-wide assay of natural radiocarbon. Science **109**, 227–228 (1949)
16. C. Bronk Ramsey, Radiocarbon dating: revolutions in understanding. Archaeometry **50**, 249–275 (2008)
17. C. Bronk Ramsey, Radiocarbon calibration and analysis of stratigraphy: the OxCal program. Radiocarbon **37**, 425–430 (1995)
18. C. Bronk Ramsey, T.F.G. Higham, P. Leach, Towards high precision AMS: progress and limitations. Radiocarbon **46**, 17–24 (2004)
19. E.C. Anderson, W.F. Libby, World-wide distribution of natural radiocarbon. Phys. Rev. **81**, 64–69 (1951)
20. H. Godwin, Half-life of radiocarbon. Nature **195**, 984 (1962)
21. P.E. Damon, J.C. Lerman, A. Long, Temporal fluctuations of atmospheric ^{14}C : causal factors and implications. Annu. Rev. Earth Planet. Sci. **6**, 457–494 (1978)
22. P.E. Damon, D. Kaimei, G.E. Kocharov, I.B. Mikheeva, A.N. Peristykh, Radiocarbon production by the gamma-ray component of supernova explosions. Radiocarbon **37**, 599–604 (1995)
23. J. Mangerud, Radiocarbon dating of marine shells, including a discussion of apparent age of recent shells from Norway. Boreas **1**, 143–172 (1972)
24. M. Stuiver, P.J. Reimer, T.F. Braziunas, High-precision radiocarbon age calibration for terrestrial and marine samples. Radiocarbon **40**, 1127–1151 (1998)
25. K. Hughen, S. Lehman, J. Southon, J. Overpeck, O. Marchal, C. Herring, J. Turnbull, ^{14}C activity and global carbon cycle changes over the past 50,000 years. Science **303**, 202–207 (2004)
26. K.A. Hughen, M.G.L. Baillie, E. Bard, J.W. Beck, C.J. Bertrand, P.G. Blackwell, C.E. Buck, G.S. Burr, K.B. Cutler, P.E. Damon, R.L. Edwards, R.G. Fairbanks, M. Friedrich, T.P. Guilderson, B. Kromer, G. McCormac, S. Manning, C. Bronk Ramsey, P.J. Reimer, R.W. Reimer, S. Remmele, J.R. Southon, M. Stuiver, S. Talamo, F. Taylor, J.v.d. Plicht, C.E. Weyhenmeyer, Marine04 marine radiocarbon age calibration, 0-26 cal kyr BP. Radiocarbon **46**, 1059–1086 (2004)
27. G. Siani, M. Paterne, M. Arnold, E. Bard, B. Metivier, N. Tisnerat, F. Bassinot, Radiocarbon reservoir ages in the Mediterranean sea and Black sea. Radiocarbon **42**, 271–280 (2000)
28. P.J. Reimer, F.G. McCormac, Marine radiocarbon reservoir corrections for the Mediterranean and Aegean seas. Radiocarbon **44**, 159–166 (2002)

29. P.A. Pirazzoli, J. Thommeret, Datation radiometrique d'une ligne de rivage a + 2,5 m pres de Aghia Roumeli, Crete, Grece. Comptes Rendus de l'Académie des Sciences Paris **284**, 1255–1258 (1977)
30. Y. Thommeret, J. Thommeret, J. Laborel, L.F. Montaggioni, P.A. Pirazzoli, Nouvelles donnees sur les rivages souleves de l'Holocene dans l'ouest de la crete. Oceanis **7**, 473–480 (1981)
31. P.A. Pirazzoli, J. Thommeret, Y. Thommeret, J. Laborel, L.F. Montaggioni, Crustal block movements from Holocene shorelines: Crete and Antikithera (Greece). Tectonophysics **86**, 27–43 (1982)
32. D. Kelletat, L. Zimmermann, Verbreitung und Formtypen rezenter und subrenzer Gesteinsbildungen an den Kusten Kretas. Essener Geographische Arbeiten **23**, 163 (1991)
33. S. Price, T. Higham, L. Nixon, J. Moody, Relative sea-level changes: reassessment of radiocarbon dates from Sphakia and west Crete. Annu. British Sch. Athens **97**, 171–200 (2002)
34. B. Shaw, N.N. Ambraseys, P.C. England, M.A. Floyd, G.J. Gorman, T.F.G. Higham, J.A. Jackson, J.-M. Nocquet, C.C. Pain, M.D. Piggott, Eastern Mediterranean tectonics and tsunami hazard inferred from the AD 365 earthquake. Nat. Geosci. **1**, 268–276 (2008)
35. L.F. Robinson, G.M. Henderson, N.C. Slowey, U-Th dating of marine isotope stage 7 in Bahamas slope sediments. Earth Planet. Sci. Lett. **196**, 175–187 (2002)
36. H. Cheng, R.L. Edwards, J. Hoff, C.D. Gallup, D.A. Richards, Y. Asmerom, The half lives of uranium-234 and thorium-230. Chem. Geol. **169**, 17–33 (2000)
37. C. Linnaeus, Tomus i. systema naturae per regna tria naturae, secundum classes, ordines, genera, species, cum characteribus, differentiis, synonymis, locis. Edito decima, reformata. Holmiae (Laurentii Salvi) **1**(-4), 1–824 (1758)
38. C. Yonge, Adaptation to rock boring in Botula and Lithophaga (Lamellibranchia, Mytilidae) with a discussion on the evolution of this habit. Q. J. Microsc. Sci. **96**, 383–410 (1955)
39. V. Jaccarini, H. Bannister, The pallial glands and rock boring in Lithophaga lithophaga. J. Zool. **154**, 397–401 (1968)
40. F. Antonioli, M. Oliverio, Holocene sea-level rise recorded by a radiocarbon-dated mussel in a submerged speleothem beneath the Mediterranean sea. Q. Res. **45**, 241–244 (1996)
41. K.R. Ludwig, *Isoplot 3.00: A Geochronological Toolkit for Microsoft Excel* (Berkeley, 2003)
42. M. Siddall, E. Rohling, A. Almogi-Labin, C. Hemleben, D. Meischner, I. Schmelzer, D. Smeed, Sea-level fluctuations during the last glacial cycle. Nature **423**, 853–858 (2003)
43. S.E. Bassett, G.A. Milne, J.X. Mitrovica, P.U. Clark, Ice sheet and solid Earth influences on far-field sea-level histories. Science **309**, 925–928 (2005)
44. E.J. Rohling, K. Grant, M. Bolshaw, A.P. Roberts, M. Siddall, C. Hemleben, M. Kucera, Antarctic temperature and global sea-level closely coupled over the past five glacial cycles. Nat. Geosci. **2**, 500–504 (2009)
45. D.P. Gillikin, A. Lorrain, S. Bouillon, P. Willenz, F. Dehairs, Stable carbon isotopic composition of Mytilus edulis shells: relation to metabolism, salinity, $\delta^{13}C_{DIC}$ and phytoplankton. Org. Geochem. **37**, 1371–1382 (2006)
46. D.P. Gillikin, A. Lorrain, L. Meng, F. Dehairs, A large metabolic carbon contribution to the $\delta^{13}C$ record in marine aragonitic bivalve shells. Geochim. Cosmochim. Acta **71**, 2936–2946 (2007)
47. W.S. Broecker, J.L. Kulp, The radiocarbon method of age determination. Am. Antiq. **22**, 1–11 (1956)
48. F.W. Shotton, An example of hard-water error in radiocarbon dating of vegetable matter. Nature **240**, 460–461 (1972)
49. E.S. Deevey, H.S. Gross, G.E. Hutchinson, G.E. Hutchinson, G.E. Hutchinson, The natural C-14 content of materials from hard-water lakes. Proc. Natl. Acad. Sci. **40**, 285–288 (1954)
50. J. Toggweiler, K. Dixon, K. Bryan, Simulations of radiocarbon in a coarse-resolution world ocean model, 2 Distributions of bomb-produced ^{14}C. J. Geophys. Res. **94**, 8243–8264 (1989)
51. K. Kleemann, Associations of corals and boring bivalves since the late Cretaceous. Facies **31**, 131–139 (1994)

52. M. Owada, Functional morphology and phylogeny of the rock-boring bivalves Leiosolenus and Lithophaga (Bivalvia: Mytilidae: A third functional clade). Marine Biol. **150**, 853–860 (2006)
53. M. Goldsworthy, J. Jackson, J. Haines, Continuity of active fault systems in Greece. Geophys. J. Int. **148**, 596–618 (2002)
54. R.E.L. Collier, D. Pantosi, G. d'Addezio, P.M. De Martini, E. Masana, D. Sakellariou, Paleoseismicity of the 1981 Corinth earthquake fault: seismic contribution to extensional strain in central Greece and implications for seismic hazard. J. Geophys.Res. **103**, 30001–30019 (1998)
55. D. Pantosti, P.M. De Martini, I. Koukouvelas, L. Stamatopoulos, N. Palyvos, S. Pucci, F. Lemeille, S. Pavlides, Palaeoseismological investigations of the Aegion fault (Gulf of Corinth, Greece). Comptes Rendus Geosci. **336**, 335–342 (2004)
56. D. Pantosti, P.M. De Martini, D. Papanastassiou, F. Lemeille, N. Palyvos, G. Stavrakakis, Palaeoseismological trenching across the Atalandi fault (Central Greece): evidence for the ancestors of the 1894 earthquake during the Middle Ages and Roman times.. Seismol. Soc. Am. **94**, 531–549 (2004)
57. A. Avallone, P. Briole, A.M. Agatza-Balodimou, H. Billiris, O. Charade, C. Mitsakaki, A. Nercessian, K. Papazissi, D. Paradissis, G. Veis, Analysis of eleven years of deformation measured by GPS in the Corinth Rift Laboratory area. Comptes Rendus Geosci. **336**, 301–311 (2004)
58. C. Hollenstein, M. MŸller, A. Geiger, H.-G. Kahle, Crustal motion and deformation in Greece from a decade of GPS measurements, 1993–2003. Tectonophysics **449**, 17–40 (2008)
59. A.N. Dia, A. Cohen, S, K., O.R., J.A. Jackson, Rates of uplift investigated through ^{230} Th dating in the Gulf of Corinth (Greece). Chem. Geol. **138**, 171–184 (1998)
60. P.M. De Martini, D. Pantosti, N. Palyvos, F. Lemeille, L. McNeill, R. Collier, Slip rates of the Aigion and Eliki faults from uplifted marine terraces, Corinth Gulf, Greece. Comptes Rendus Geosci. **336**, 325–334 (2004)
61. L.C. McNeill, R.E.L.I. Collier, Uplift and slip rates of the eastern Eliki fault segment, Gulf of Corinth, Greece, inferred from Holocene and Pleistocene terraces. J. Geol. Soc. Lond. **161**, 81–92 (2004)
62. P. Clarke, R. Davies, P. England, B. Parsons, H. Billiris, D. Paradissis, G. Veis, P. Denys, P. Cross, V. Ashkenazi, R. Bingley, Geodetic estimate of seimsmic hazard in the Gulf of Korinthos. Geophys. Res. Lett. **24**, 1303–1306 (1997)
63. M. Goldsworthy, J. Jackson, Migration of activity within normal fault systems: examples from the Quaternary of mainland Greece. J. Geol. Soc. **157**, 967–981 (2000)
64. N. Mouyaris, D. Papastamatiou, C. Vita-Finzi, The Helice fault. Terra Nova **4**, 124–129 (1992)
65. S.C Stiros, M. Arnold, P.A Pirazzoli, J. Laborel, F. Laborel, S. Papageorgiou, Historical coseismic uplift on Euboea Island, Greece. Earth Planet. Sci. Lett. **108**, 109–117 (1992)
66. C. Vita-Finzi, Evaluating the Quaternary uplift in Greece and Cyprus. Geol. Soc. Lond. Special Publ. **76**, 417–424 (1993)
67. B. Keraundren, C. Faulguères, J.J. Bahain, D. Sorel, Y. Yokoyama, Nouvelles datations radiométriques des terraces marines de Corinthie (Péleponnèse septentrional, Grèce). Comptes Rendus de l'Académie des Sciences Paris **320**, 483–489 (1995)
68. I. Stewart, Holocene uplift and palaeoseismicity on the Eliki Fault, Western Gulf of Corinth, Greece. Ann. Geophys. **39**, 575–588 (1996)
69. I. Stewart, C. Vita-Finzi, Coastal uplift on active normal faults: the Eliki Fault, Greece. Geophys. Res. Lett. **23**, 1853–1856 (1996)
70. P.A. Pirazzoli, S.C. Stiros, M. Arnold, J. Laborel, F. Laborel-Deguen, Late Holocene coseismic vertical displacements and tsunami deposits near Kynos, Gulf of Euboea, Central Greece. Phys. Chem. Earth. Part A: Solid Earth Geodesy **24**, 361–317 (1999)
71. P.A. Pirazolli, S.C. Stiros, M. Fontugne, M. Arnold, Holocene and Quaternary uplift in the central part of the southern coast of the Corinth Gulf (Greece). Marine Geol. **212**, 35–44 (2004)

Chapter 5
Geomorphology

5.1 Preface

The bulk of this thesis focuses on the tectonics of the Hellenic subduction zone, using the historical record, radiometric dating and seismology to investigate how present-day faulting accommodates the relative motion between Nubia and the Aegean. Additionally, a pilot study carried out to attempt to investigate the potential of using topography and geomorphology to constrain the long-term uplift pattern of western Crete is described here. Although the results are very much preliminary, and further work is required to draw firm conclusions, I include this study to show the limitations of the techniques that were investigated, and to assist with any further work in this area.

5.2 Introduction

In Chap. 2, I described evidence for uplift of the western part of Crete by up to 10 m during the earthquake of AD 365. The AD 365 paleo-shoreline is the only such feature on the cliff-faces of Crete, suggesting that any similar previous earthquakes occurred prior to the stabilisation of sea-level following the last glacial maximum. Nevertheless, there is abundant evidence for sustained rapid uplift on Crete.

Above the AD 365 shoreline, the coastal limestone outcrops are perforated by the boreholes of *L. lithophaga* to an elevation of at least 50 m in south-west Crete (see Chap. 9). Although some shells are preserved in the boreholes, they are well-cemented with calcareous dripstone making them difficult to sample.

In addition, marine terraces can be recognised in the west of the island; unfortunately these terraces tend to be isolated and of limited extent. Nevertheless, they can be used to estimate an uplift rate for the west of Crete by radiometric dating of marine fauna, and by matching terrace elevations and dates to the eustatic sea-level curve. This is the primary motivation for the fieldwork that I describe in the rest of this chapter.

B. Shaw, *Active Tectonics of the Hellenic Subduction Zone*, Springer Theses,
DOI: 10.1007/978-3-642-20804-1_5,

Further evidence for the long-term uplift of western Crete comes from a series of rivers on the south coast that have been forced to incise as their baseline (i.e. relative sea-level) has fallen, resulting in deep, narrow gorges. At Frangokastelli (see Fig. 1.2 for locations) in south-central Crete, rivers incise into three generations of uplifted fan deposits (see Chap. 9 and [1, 2]). The Samaria gorge, incising through the Levka Ori, is just 1 m wide at its narrowest point and several hundred metres deep in places.

Finally, Miocene and Late Pliocene marine sediments can be recognised at elevations of hundreds of metres above sea-level. Using the assumption that Oligocene and Lower Miocene nappe units were horizontal in the Middle Miocene, and measuring the difference in altitude between these units in the Aegean Sea and in the mountains of western Crete today, Le Pichon and Angelier [3] estimate that a lower bound on the average regional uplift rate since the middle Miocene is 0.3–0.4 mm/year.

The long-term uplift of Crete is thought to arise from underplating by subducted sediment [3]. In spite of the extreme thickness (up to 10 km) of sediment on the African plate, isotopic analysis of Cycladic magmas suggests that very little sediment is subducted to depths at which melting occurs beneath the volcanic arc [4–6]. The incoming sediment must therefore detach from the subducting slab, either by scraping off to form the growing accretionary prism, or by underplating beneath the overriding Aegean. Slip along splay faults, as I suggest occurred in AD 365, is an efficient means of achieving this. If the Hellenic Trench marks the surface expression of such a splay fault (see Chap. 2), then the south-west corner of Crete should be uplifting most rapidly, with uplift rate decreasing to the north and to the east.

There are several possible approaches to investigating the long-term uplift pattern. The most obvious is to determine whether the first-order morphology of western Crete matches the pattern of uplift seen in the AD 365 earthquake. An alternative approach would be to map out the profiles of river networks in the area. Rivers must downcut in response to the rapid uplift of land with respect to sea-level. If the position of the baseline (i.e. sea-level) varies through time, as a result of uplift or due to eustatic sea-level variations, a knickpoint may be cut. With time, the knickpoint will migrate up the river profile. The positions of knickpoints can be used to infer spatial and temporal uplift histories (e.g. Roberts and White [7]). Accurate profiles of river gorges may therefore be used to constrain the uplift history of an area.

The interaction between eustatic sea-level variations and uplift of the land also leads to the formation of marine terraces. When sea-level is stationary with respect to the land, a wave-cut platform is formed. If relative sea-level then drops, this platform may be preserved as a terrace, provided it is lifted above the height of the next highstand. The elevations of terraces can therefore be used, along with a reliable sea-level curve, to estimate the uplift-rate of the land.

5.2.1 *Kinematic GPS*

All of the above approaches rely on accurate knowledge of the topography at small scales. Digital elevation models can be downloaded at 90 m (SRTM) and 30 m (ASTER), but it is possible to construct DEMs of extremely high accuracy (cm scale) using kinematic GPS. I therefore carried out kinematic GPS surveys of uplifted and tilted surfaces and of river profiles in western Crete. The technique involves comparing the GPS signal recorded at a stationary base station, which is a receiver mounted on a fixed tripod, with that recorded by one or more roving GPS receivers. The roving receivers are carried at a fixed height above ground level across the area being surveyed. Satellite positions are recorded every second, and the data collected by the base is used to correct the rover data, resulting in extremely accurate (cm scale) relative locations and elevations. The surveys were carried out with the help of Gareth Roberts and Richard Kahle and the results are described in the following sections.

5.3 Large-Scale Morphology of Crete

If the uplift of western Crete is controlled by slip along a fault surfacing at the Hellenic Trench, as was inferred for the AD 365 earthquake, then the topography of western Crete might be expected to reflect this pattern of uplift, with the highest elevations in the south-west of the island, and decreasing elevations to the north and to the east.

A swath cross-section perpendicular to the Hellenic trench confirms that the topography of Western Crete is asymmetrical, with the steeper slopes on the south coast, closest to the Hellenic Trench, and gentler slopes to the north (Fig. 5.1). In this figure, the points that were included in the swath (within 2 km of the line of section) are plotted on the inset map. Black crosses mark the elevations of these points in the cross-section. The red line shows the mean topographic elevation along the line of section. The horizontal scale, or wavelength, of this ramp appears to have a similar scale to the pattern of uplift following the AD 365 earthquake and the base of the inferred AD 365 fault coincides approximately with the north coastline of Crete (see Chap. 2). In the AD 365 earthquake, maximum uplift was 10 m, and this uplift decreased to zero over a distance of 50 km. Today, the tilt of the north coast is $\sim 1^\circ$, a tilt that could be achieved in 100 AD 365 type earthquakes, or in 500,000 years, if the estimates of their repeat time from Chap. 2 are approximately correct.

Although the NE-SW asymmetry, expected if uplift of Western Crete is controlled by a fault surfacing at the Hellenic Trench, is observed to an extent, the highest mountains are the Levka Ori in central western Crete (see Fig. 5.2). Here, competent limestone of the Mesozoic Tripolis nappe outcrops. In contrast, the topographically lower south-west corner is dominated by more friable, finely bedded phyllites and quartzites [8]. The pattern of uplift is therefore strongly modified by the competence of the uplifted geological units to produce the topography observed today. In addition to the geological control, recent N-S normal faults are responsible for the position of the N-S peninsulas in western Crete [9] and also modify the large-scale topography (Fig. 5.2). The large-scale topography alone, therefore, cannot be relied on to reveal a clear pattern of long-term uplift in western Crete.

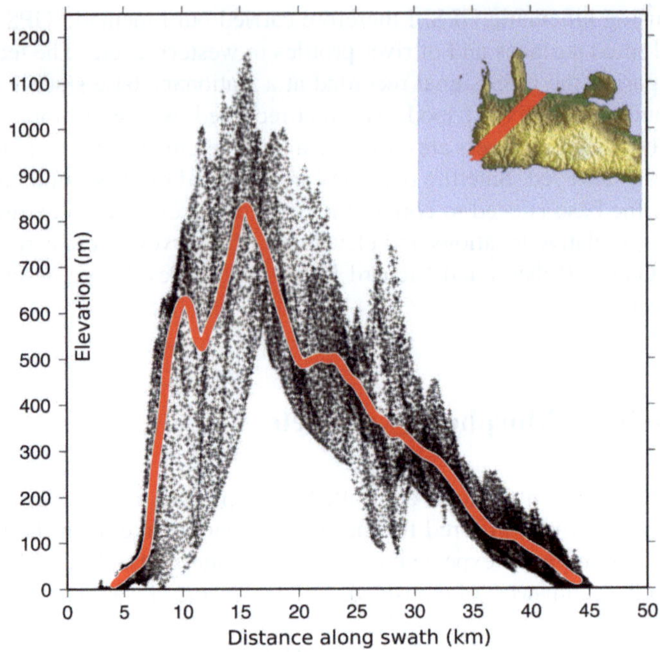

Fig. 5.1 Swath cross-section through western Crete, using the 90 m SRTM DEM. The *inset* map shows the points that were incorporated in the 4 km wide swath-section (*red dots*). The elevation of these points are plotted in the main panel as *black crosses*. The *red line* represents the mean of these points

The north coast of western Crete is dominated by Neogene deposits, principally marly limestones with sandy and silty intercalations. The general dip of these deposits is to the north/north-east, although Andronopoulos [8] suggests there is a large scatter in strike and dip in this area. Here, the interfluves slope gently to the north-east, and appear possibly to be the remnants of an eroded, once-horizontal surface that has been uplifted and tilted since its formation. The surface does not parallel the bedding at the limited number of points where strike, dip and rake could be measured. Nor is the top of the surface defined by a marine cap, such as that observed on the terraces on the south coast described later. Figure 5.3 is a photograph taken from the top of an interfluve, looking NW towards the adjacent interfluve. The vegetated river valley can be seen in the foreground.

To investigate whether the interfluves do have a consistent dip direction, it is possible to construct a surface through the tops of the spurs. The cartoon in Fig. 5.4 explains how the points on top of the interfluves were automatically extracted in order for a surface to be constructed through them. The original DEM (in this case the ASTER DEM) was smoothed using a Gaussian filter, initially of width 5 km. Points whose elevations were higher in the original DEM than in the smoothed

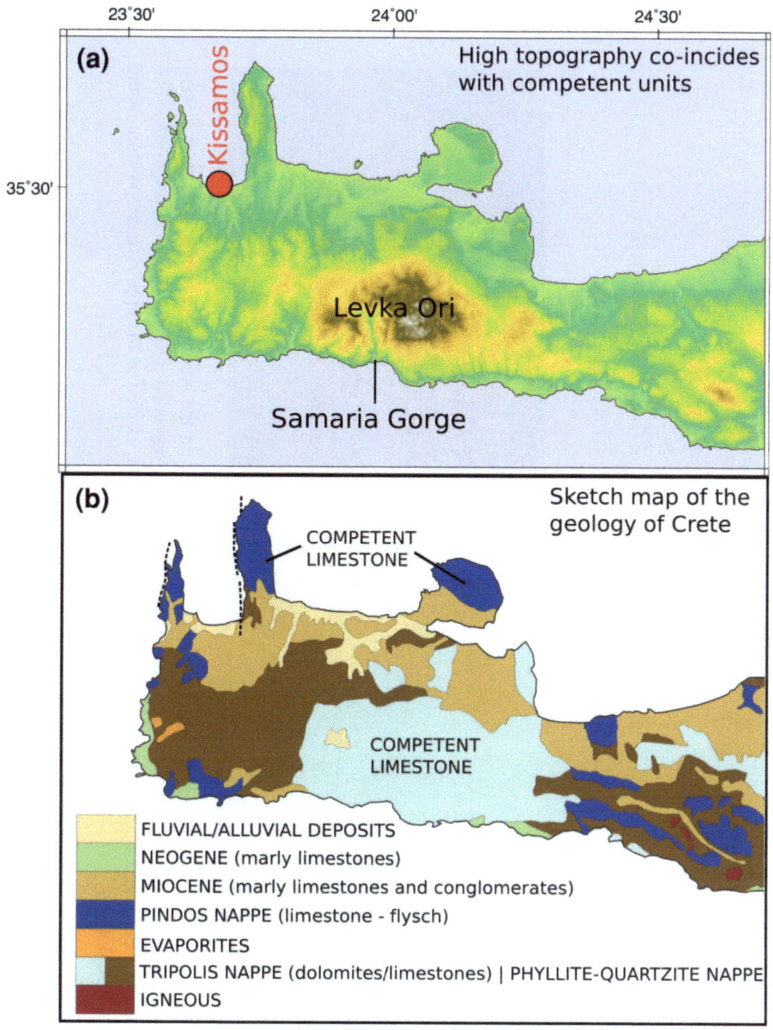

Fig. 5.2 Topographic (**a**) and geological (**b**) maps of western Crete. The areas of highest topography in western Crete coincide with the areas where competent limestones outcrop at the surface, suggesting that the pattern of uplift is preserved differentially depending on the geology. *Dashed lines* in (**b**) show the approximate position of significant N–S normal faults in the west of Crete, from Armijo et al. [9]. The geological map is a simplified version of the 1:50,000 Geological Map of Greece, published by the Institute of Geology and Mineral Exploration in 1993

version were selected, and the process repeated using progressively smaller filter widths until only points at the tops of the interfluves remained.

The result of fitting a high-tension surface to interfluve points selected in this way is shown in Fig. 5.5. In Fig. 5.5a, the points used to construct the surface are shown as black dots on the ASTER DEM (contoured at 50 m intervals). Figure 5.5b shows

Fig. 5.3 This mosaic of photographs (*left panel*) shows the gentle tilt to the NE of the interfluves in the area of Kissamos. The photos were taken looking NW from the adjacent interfluve (*black arrow* labelled *A* in *right panel*) and show the vegetated river valley in the foreground. The *right panel* is an image is taken from Google Earth and again shows the gentle tilt of the interfluves, as well as the approximate position of the lithological boundary between marl and limestone

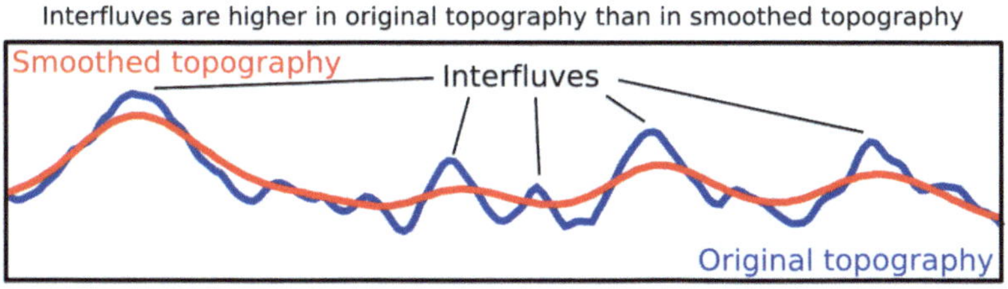

Fig. 5.4 This cartoon explains how the points on top of the interfluves were automatically extracted in order for a surface to be constructed through them. The original DEM (*blue*) was smoothed using a Gaussian filter, initially of width 5 km, to produce the smoothed topography (*red*). Points whose elevations were higher in the original DEM than in the smoothed version were selected, and the process repeated using progressively smaller filter widths until only points at the tops of the interfluves remained

the approximate boundary between limestone and marl. This geological boundary correlates strongly with the topographic slope. A cross-section through the interfluve-derived surface (Fig. 5.5c, d) confirms the strong geological control on the topography: the break in slope coincides with the lithological boundary. In the area of the cross-section, the dip of the marl surface is to the NE, as observed in the field, but further to the east the slope is to the NW, away from the high limestone topography. Using this technique, there is very little evidence for a continuous tilted surface within the marl and the slope of the topography seems to be controlled by the position of outcropping limestone.

An alternative approach is to plot cross-sections through the topography where the marl outcrops, in a direction perpendicular to the dip direction of the interfluves to see whether they form a continuous tilted surface. Figure 5.6 shows the results of plotting cross-sections through the topography. The white line on the map, and the black dots on the cross-sections indicate the results of a kinematic GPS survey carried out over this area. Grey dashed lines in the cross-sections are horizontal. From these cross-sections, the interfluves appear to form a fairly continuous tilted surface over a limited area. This surface is consistent with a planar tilt of $\sim 4°$.

Therefore both the overall topography of western Crete, and, on a smaller scale, the topography in the area of Kissamos, appears to be so strongly influenced by the competence of the outcropping lithologies that the pattern of long-term uplift expected from repeated slip along a fault surfacing at the Hellenic Trench is difficult to distinguish, if it exists. The following sections focus on smaller-scale features, including a discussion of the suitability of kinematic GPS surveys for mapping river profiles and constructing DEMs of terraces.

5.4 Rivers

Rivers respond to rapid uplift by downcutting; if uplift rate varies with time, knickpoints (changes in gradient) will be formed that will migrate up the river profile. In order to recognise knickpoints, accurate river profiles are required. It is often not

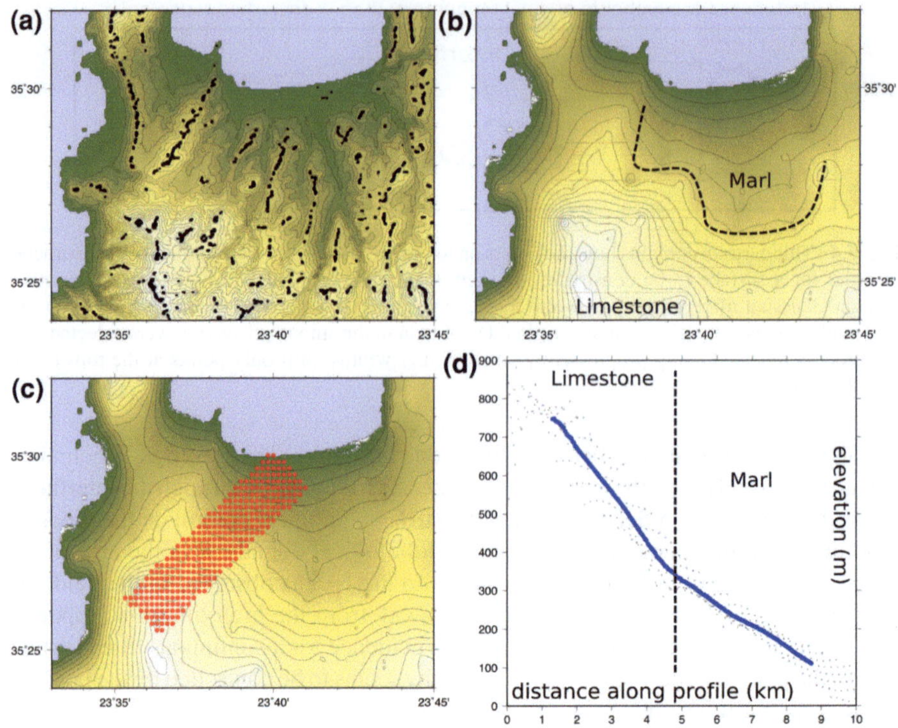

Fig. 5.5 **a** ASTER DEM contoured every 50 m. The *black dots* are points on the interfluves that were used to construct a surface. These points were identified by smoothing the DEM, using a Gaussian filter of width 5 km, and then picking the points whose elevations were higher in the original DEM than in the smoothed version. The points are those remaining after four iterations with progressively smaller smoothing widths, and lie along the tops of the most prominent interfluves. **b** Smooth surface fitted through these points, using a cubic interpolation algorithm with high tension. The surface is also constrained at the coastline by fitting points with an elevation of zero. **c** Points used in a swath-section as *red dots*. **d** Resulting cross-section. The points used are plotted as *black crosses* and the mean is shown by the *blue line*

possible to obtain such profiles from satellite-derived topography (such as SRTM), especially where rivers flow through gorges and slot canyons, precisely where incision is occurring and knickpoints are to be expected, since the satellite topography is strongly biased towards the upper surfaces of these features.

The primary focus of this section is to establish whether kinematic GPS can be used to construct river profiles by walking along the river bed. Figure 5.7 shows the positions of the rivers that were mapped using kinematic GPS. The first river mapped was on the northern side of Western Crete. A map view of this river, and of the GPS data-points collected during this survey, is plotted in Fig. 5.7b, where the blue dots are locations of points recorded. The topography on which the GPS data is plotted comes from the SRTM DEM. Most of the blue points follow the river valley closely,

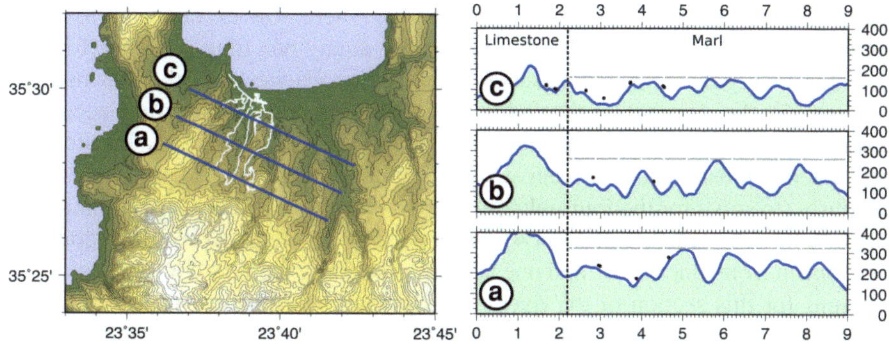

Fig. 5.6 Map showing the path taken during a kinematic GPS survey of the area (*white line*) and the positions of three cross-sections through the interfluves (*blue lines*). **a–c** Cross-sections through topography (*blue lines*) as plotted on the map. *Grey dashed lines* are horizontal. The *vertical black line* shows the approximate boundary between limestone and marl. The *black dots* are the positions of points recorded during the kinematic GPS survey

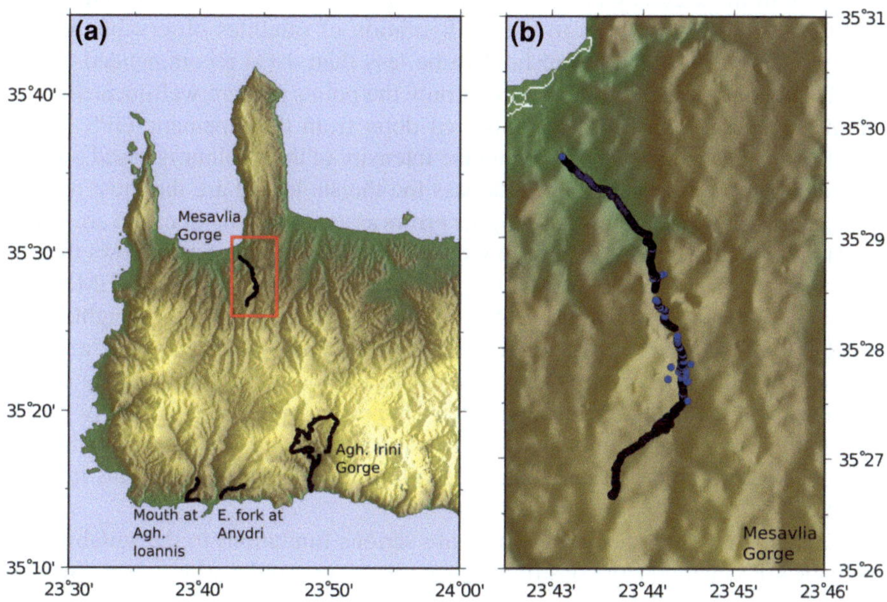

Fig. 5.7 **a** Map of western Crete showing the four rivers that were mapped using kinematic GPS. **b** GPS data for the Mesavlia Gorge [highlighted by the red box in (**a**)] are plotted as *blue dots* on the SRTM DEM

but in some areas the points show a large scatter (more than 200 m) due to poor GPS reception.

The photographs and profile in Fig. 5.8 illustrate the difficulties encountered when carrying out the survey. At point 1, in the upper course of the river, the channel is open, surrounded by olive and citrus orchards. There was excellent GPS coverage at this point, with up to 10 satellites visible at any one time, and therefore the locations are highly accurate (cm scale) in this region. However, further downstream (points 2–5), the river passes through narrow channel constrictions, often accompanied by apparent knickpoints (changes in gradient). In order to resolve a location, the GPS antenna must be able to receive signals from a minimum of 4 satellites simultaneously (to solve for the four unknowns: x, y, z and time). Where the channel was narrow and steep-sided (e.g. point 4), often only three satellites (or fewer) could be seen: one in front, one above, and one behind, explaining the very poorly constrained locations for this section of the river. The kinematic GPS survey therefore cannot accurately resolve the profile in this region. Three more river profiles were mapped using kinematic GPS, and the results are presented in Chap. 10. In all cases, channel constrictions and dense vegetation cover prevented contact with sufficient satellites for accurate profiles to be constructed along the entire length of the river.

Although the profile of the river could not be determined along its entire length, it is possible to plot only the points with recorded errors below an acceptable threshold. A measure of the accuracy of each data-point (PDOP) is recorded by the GPS units during the survey and relates to the configuration of satellites observable at each point. By setting the PDOP threshold to be less than 4 (as recommended by the GAMIT manual), it is possible to extract only the points that are well-located.

Figure 5.9a is a plot of data-points (red dots) from the kinematic GPS survey along this river. The dots are scaled and the intensity of their colour is based on their recorded errors, so that large red dots pass the threshold and are the most reliable points; the smaller pink points have higher errors associated with them and are scaled according to their reliability. The map view of these points (Fig. 5.9b) shows that the points with the lowest errors plot within the river valley in the ASTER DEM.

The yellow and blue dots in Fig. 5.9a represent the corresponding heights from the SRTM and ASTER DEMs respectively for each point of the survey. They have been scaled and shaded in the same manner as, and based on the accuracy of, the GPS survey points. A static elevation offset of approximately 20 m is clearly visible between the survey data and the DEMs. This is to be expected since kinematic GPS techniques record very accurate relative locations, but the absolute heights must be calibrated.

This field-study has shown that there are serious limitations to the suitability of kinematic GPS methods for constructing accurate river profiles where river channels are narrow, or where there is dense vegetation. By walking down the rivers, it was possible to identify sudden changes in gradient in the field, but in these conditions it is not possible to identify knickpoints using kinematic GPS.

5.5 Terraces

When sea-level remains stationary with respect to the land, the sea erodes landward forming a wave-cut platform. In order for a terrace to be preserved, this platform must be lifted out of the reach of future highstands. The most recent major highstand

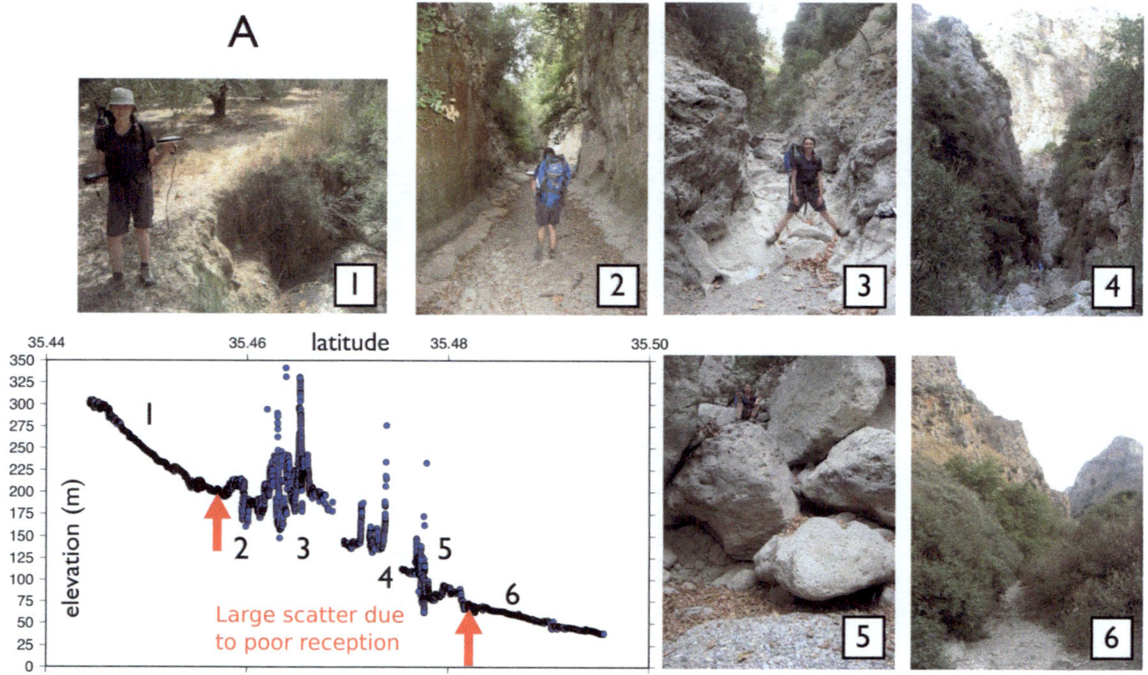

Fig. 5.8 Photographs and profile of the first river mapped (Mesavlia Gorge; see Fig. 5.7). The photographs were taken at progressively lower elevations. Between the *red arrows*, the scatter is very large (around 250 m in places) due to poor satellite reception, so the river profile cannot be resolved here even though rapid local changes in gradient (knickpoints) could be detected in this region by walking down the river bed

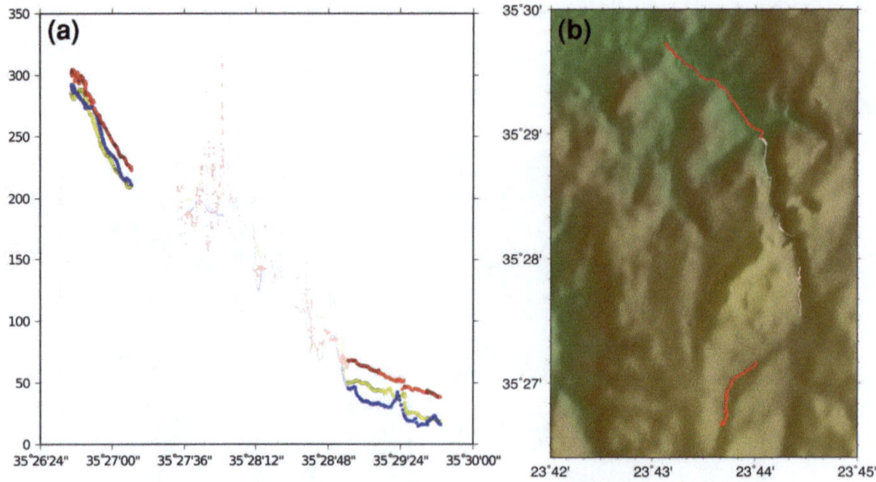

Fig. 5.9 **a** Data points (*dots*) collected during the kinematic GPS survey along the river, plotted on the ASTER DEM. The *dots* are scaled and the intensity of colour is based on their recorded errors, so that *large red dots* pass the threshold (see text) and are the most reliable points; the *smaller pink points* have higher errors associated with them and are scaled according to their reliability. The points with the lowest errors plot within the river valley in the ASTER DEM. **b** Plot of the heights of these points in metres against latitude; the *dots* are scaled and shaded as in (**a**). The *yellow* and *blue dots* represent the corresponding heights from the SRTM and ASTER DEMs respectively for each point of the survey. They have been scaled and shaded in the same manner, based on the accuracy of the GPS survey points

is the MIS 5e highstand (125,000 year BP), although there have been a number of more minor highstands since then. In order for any of the intermediate highstands to be preserved, uplift must be extremely rapid, usually more than 1 mm/year (see Armijo et al. [10] and Sect. 5.5).

Uplifted marine surfaces can be recognised above the AD 365 shoreline at a number of sites in western Crete; the clearest terraces are found at Paleochora and Chrisoskalitissa on the south coast (Fig. 5.10 and Chap. 2). To determine accurate elevations and morphology of the terraces, Gareth Roberts, Richard Kahle and I surveyed the surfaces using kinematic GPS, and constructed DEMs of the uplifted surfaces. The absolute terrace heights were calibrated by walking down to sea-level.

The clearest terrace on the south coast of Crete is above the harbour at Paleochora. The photographs in Fig. 5.11 show the appearance of the terrace and the surfaces that were surveyed. The bedrock is steeply dipping towards the left of the photograph in Fig. 5.11c and is unconformably overlain by a recent consolidated marine cap, containing abundant marine bivalves. The path taken during the kinematic GPS mapping of the terrace, and the resulting DEM are shown in Fig. 5.11d, e. The top surface of the terrace is flat and well constrained by the GPS survey, as the base station and roving receiver retained excellent connections with up to ten satellites at

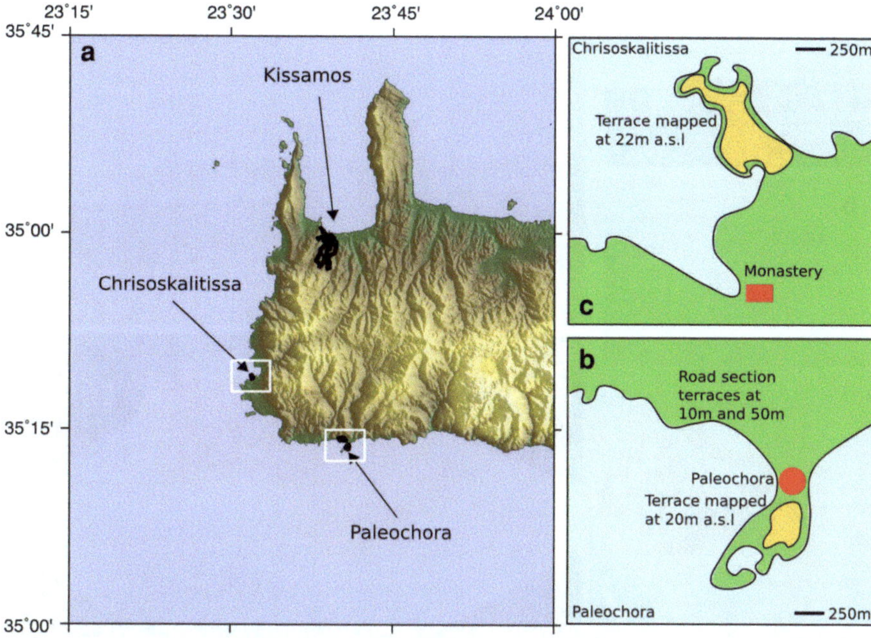

Fig. 5.10 **a** Uplifted marine surfaces at Chrisoskalitissa, Paleochora and Kissamos. The positions of terraces mapped using kinematic GPS at Chrisoskalitissa and Paleochora are highlighted by *white boxes*. **b**, **c** Sketch maps of these terrace surfaces

Table 5.1 ^{14}C dates of samples from terraces in western Crete (identified by Oxford Radiocarbon Accelerator Unit sample number)

Ox ID	Organism	Location	Height (m)	Top of paleo-shore (m)	^{14}C age BP	1σ Range years	2σ Range years
13809	bivalve	CHS T1	20–24	7.9	39180 ± 300	41101–40661	41851–40327
14085	bivalve	CHS T2	20–24	7.9	37700 ± 250	40615–39981	40649–39237
16995	bivalve	PAL T	20–24	9.0	47300 ± 550	53063–45967	53461–45621

The 1σ and 2σ ages for the samples are years BC, and were obtained by comparing the radiocarbon ages with the ^{14}C record from the Cariaco Basin [14] with a 400 year subtraction made from the conventional ages to account for the marine reservoir effect.

all times during the survey. The marine cap is at an elevation of 20 m above present-day sea-level and is the surface dated at 47,300 ± 550 year. BP (PALT in Table 5.1).

To the west of Paleochora (35° 14.262′ N 23°39.142′ E) is a series of dramatic marine notches, terraces and sea-caves complete with *Lithophaga lithophaga* borings. Figure 5.12 shows the lower terrace. A cross-section through the DEM of the two terraces, produced using kinematic GPS, shows that the lower lies at 10 m above sea-level (1–1.5 m above the AD 365 shoreline) and the higher surface is at an el-

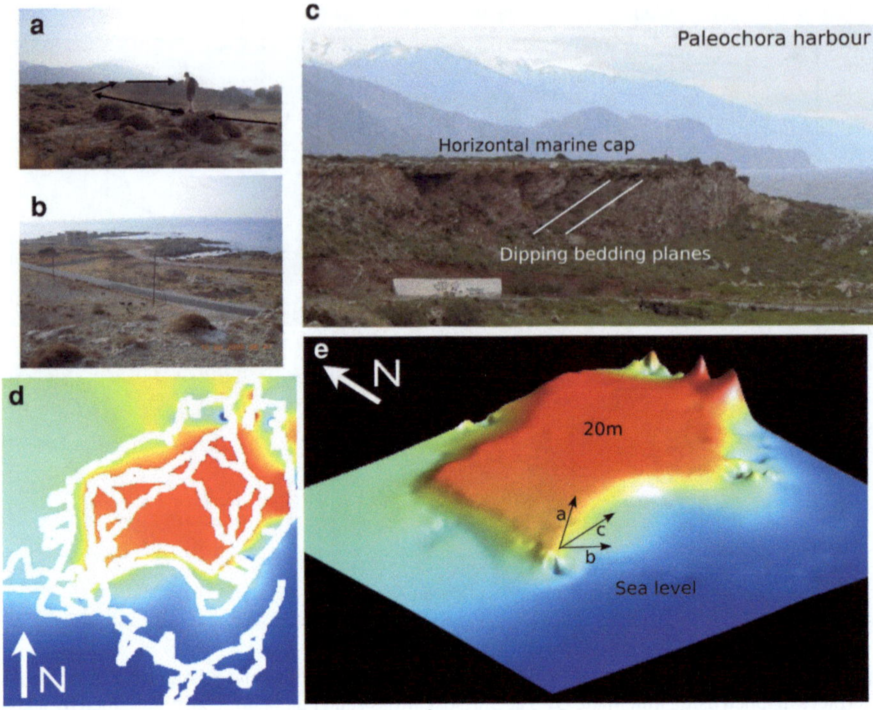

Fig. 5.11 20 m terrace at Paleochora harbour. **a** Photograph showing the extensive flat top of the terrace. **b** The low, flat surface just above sea-level that was used to define the base of the terrace. **c** The highly tilted bedrock (tilted towards the *left* of the photograph) unconformably overlain by a recent marine cap. **d** Path walked during the survey overlain on the resulting DEM. **e** 3-D perspective view of the DEM. *Arrows* labelled *a*, *b* and *c* begin at the point from which the photos were taken and show their respective directions

evation of 50 m above sea-level. In addition to these terraces, marine caves can be recognised, inside which notches are preserved. The white arrows in Fig. 5.12c mark the positions of two clear notches, one at ~20 m, and the other perhaps 2–3 metres higher. The lower of these notches has beach pebbles cemented beneath it, revealing its erosive origin. The higher notch is marked by abundant *Lithophaga lithophaga* boreholes (see Chap. 9 for more details).

The final terrace surveyed is a 22 m high terrace at Chrisoskalitissa. Although this terrace has a clear marine carbonate cap, it has a more undulating surface and is more incised than the terrace at a similar height at Paleochora. Figure 5.13a is a mosaic of photographs showing the surface of the terrace. Figure 5.13b shows the paths taken during the survey, and the resulting DEM. The DEM shows that the surface is a continuous feature at 22 m elevation, although the terrace is not as clear as the 20 m terrace at Paleochora. At Chrisoskalitissa, bivalves from the terrace were dated at 37,700 ± 250 year BP and 39,180 ± 300 year BP (CHST1, T2 in Table 5.1)

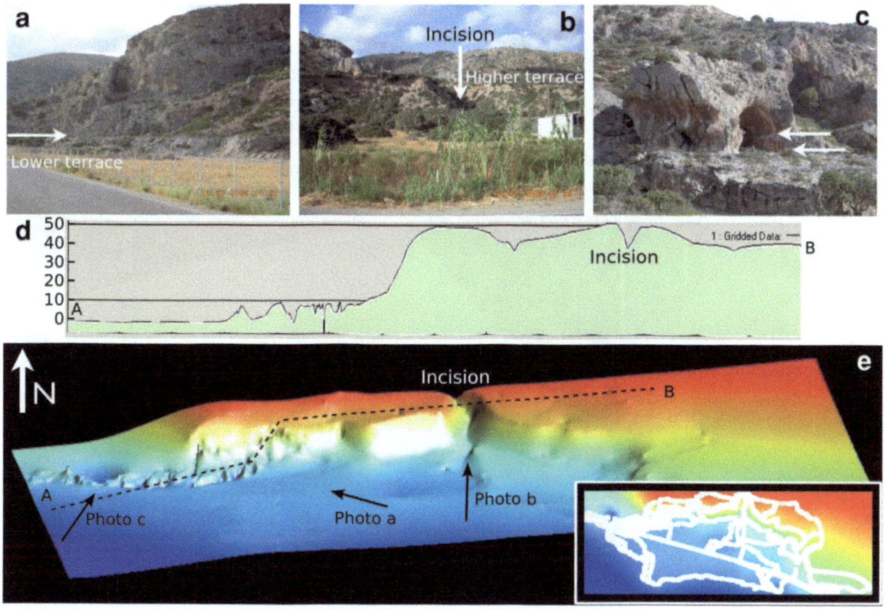

Fig. 5.12 10 and 50 m terraces at Paleochora. **a** Lower (10 m) terrace (*white arrow*). **b** Higher (50 m) terrace; the *white arrow* shows deep incision in the surface. **c** Uplifted notches and caves (at the levels of the *white arrows*). **d** Topographic profile through the DEM along the *dashed line* in (**e**) showing the elevations of the terraces (*solid lines*). **e** 3-D perspective view of the DEM produced using kinematic GPS data. *Arrows* show the location and look-direction of the photos. The *inset* shows the track coverage (*white line*) during the GPS survey overlain on the DEM

Figure 5.14 shows a map of 'flat' areas in south-west Crete based on the 30 m ASTER DEM, determined by taking the gradient of the DEM and recording the slope along the maximum dip direction. Here, 'flat' is defined as a slope of less than 1:10. A histogram, with pixel size 30 m, of elevations of such regions that are less than 100 m above sea-level is shown in Fig. 5.14b (there were no significant flat areas above 100 m elevation). Assuming that the absolute elevations are correct, flat surfaces can be recognised at elevations between 10 m and 65 m, with marked spikes at 20 m, 35 m, 40–50 m and 55–60 m. Terraces at 20 m and 50 m have been observed in the field, but it is possible that some of these flat areas are also uplifted terraces, and that there is a complicated set of small terraces between 10 m and 65 m elevation in SW Crete.

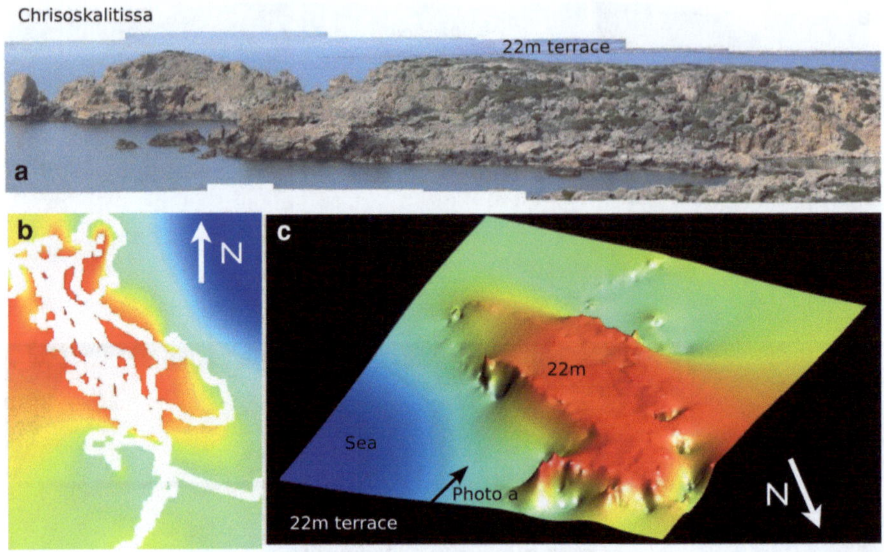

Fig. 5.13 22 m terrace Chrisoskalitissa. **a** Photograph showing the morphology of the terrace. **b** Path walked during the GPS survey overlain on the DEM. **c** 3-D perspective view of the DEM produced by the kinematic GPS survey. The *arrow* shows the direction of the photograph (**a**)

5.6 Discussion

5.6.1 Radiocarbon Dates of Terraces

Marine bivalve samples (Fig. 5.15) were collected from the 20 m terrace at Paleochora and the 22 m terrace at Chrisoskalitissa, and were radiocarbon dated at the Oxford Radiocarbon Accelerator Unit (see Chaps. 2 and 4). The resulting ages (Table 5.1) suggest that the terrace at Chrisoskalitissa formed at between BC 39,000 and BC 42,000, or 41,000–44,000 year BP. The bivalve collected from Paleochora dated from between BC 45,000 and BC 53,000, or 47,000–55,000 year BP. It is likely that the two terraces, which have a similar present-day elevation, and are found at sites that were uplifted by approximately the same amount in AD 365, were formed during the same sea-level stillstand.

An examination of the sea-level curve of Rohling et al. [11] for the last 140,000 years (Fig. 5.16) reveals that there was a complicated intermediate highstand between 60,000 and ~40,000 year BP (marine isotope stage, MIS, 3), and it is probably during this series of stillstands that the 20 m terraces at Paleochora and Chrisoskalitissa were cut. Between 40,000 and 50,000 year BP, sea-level was 75–85 m below the present-day level, suggesting an average uplift rate of 1.9–2.6 mm/year.

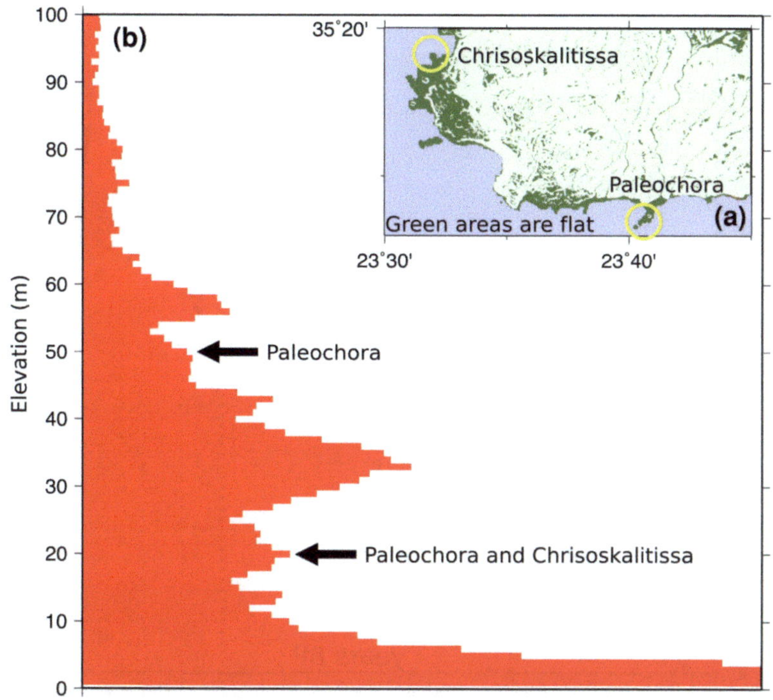

Fig. 5.14 **a** 'Flat' areas (*green*) in south-west Crete based on the 30 m ASTER DEM. Here, 'flat' is defined as a slope of less than 1:10. **b** Histogram of the elevations of such regions that are less than 100 m above sea-level. There are no substantial flat areas higher than 100 m

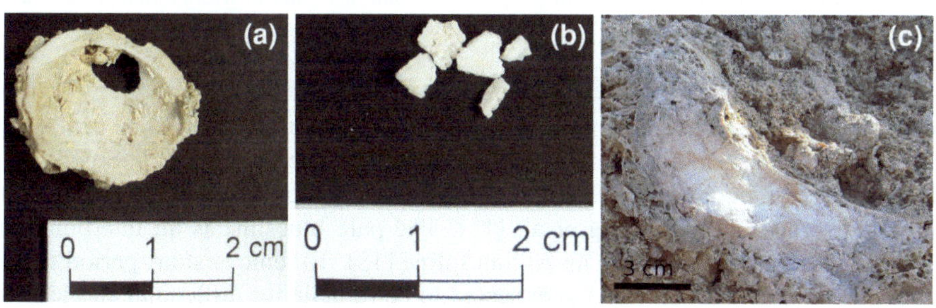

Fig. 5.15 **a** Bivalve sample collected from the 20 m terrace at Paleochora. **b** Cleaned sample obtained from (**a**) prior to dating. **c** In-situ example of the bivalves collected from the 22 m terrace at Chrisoskalitissa

5.6.2 Terrace Modelling

Assuming that terraces are cut whenever sea-level remains stationary with respect to the land, it is possible combine a sea-level curve with a range of uplift rates to forward

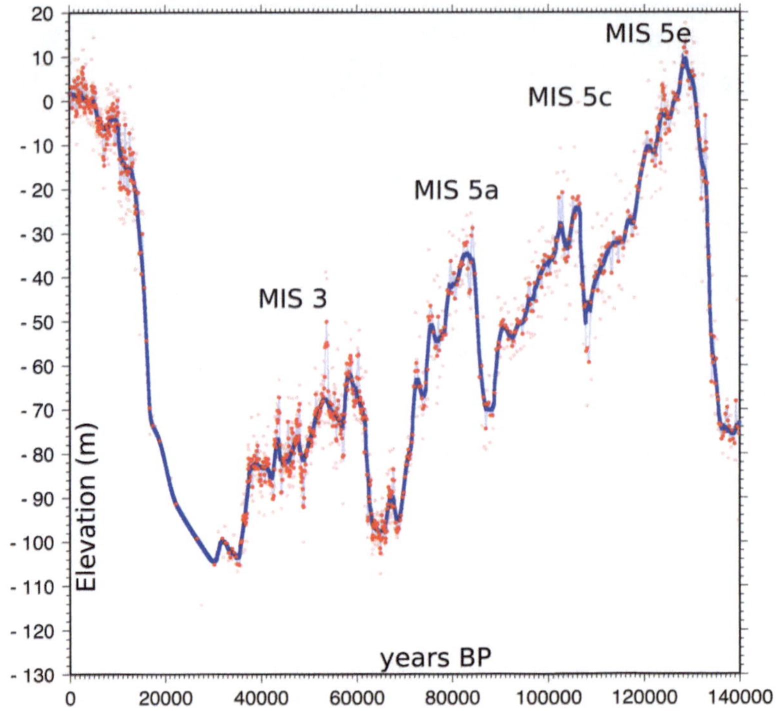

Fig. 5.16 Sea-level curve plotted for the last 140,000 years from Rohling et al. [11]. *Small pink dots* are raw data points, and the *red dots* result from their application of a three-point moving average. The *pale blue line* is my interpolation between these *red points*, using an Akima spline [13]. The *dark blue line* shows the result of applying a Gaussian filter of width 3,000 years to this interpolated curve

model the expected elevation of terraces (e.g. [10, 12]). I used the Rohling et al. [11] sea-level curve determined for the Red Sea which is plotted in Fig. 5.16. Small pink dots are raw data points while red dots result from an application of a three-point moving average, after Rohling et al. [11]. The pale blue line is an interpolation between these red points using an Akima spline [13]. To remove short-period noise from the sea-level curve, which is expected to be reliable for millennial and longer time-periods [11], I applied a Gaussian filter of width 3,000 years to this interpolated curve (dark blue line).

I calculated the position of sea-level through time relative to the land for a range of uplift rates and produced histograms showing the number of years that sea-level stood at a particular elevation. Such histograms can be combined into a single plot of uplift rate against present-day elevation (Fig. 5.17). Here, the colour palette represents the number of counts in each bin. Bright yellow/green lines represent relative stillstands, and therefore potential terrace elevations. The present-day elevation of these still-stands increases with increasing uplift rate, as expected. Figure 5.17a is constructed from the unfiltered sea-level curve. The same major terrace-cutting events can be seen when the filtered version of the sea-level curve is used Fig. 5.17b and the data is less noisy. Figure 5.17c is calculated in the same way, except that terraces are destroyed if they are submerged beneath sea-level so that only the youngest terraces are preserved at each elevation. The colour palette has been modified here to enhance the terraces.

This simple modelling relies on a number of assumptions. Firstly, terraces are assumed to form at elevations at which sea-level remains stationary for some time. Secondly, uplift is assumed to occur at a constant rate, whereas it is more likely that the uplift of western Crete occurs in large earthquakes, like the AD 365 earthquake, with a long repeat time. The final assumption is that the eustatic sea-level curve is accurate enough to justify such an analysis. Recent published sea-level curves, based on a variety of data, have the same first-order morphology, but this study is strongly dependent on the accuracy of the sea-level curve.

Based on the radiocarbon dates of bivalves from the 20 and 22 m terraces at Paleochora and Chrisoskalitissa, these terraces were formed between 40,000 and 50,000 year BP. The yellow horizontal line in Fig. 5.14b shows the range of uplift rates that would produce terraces of this age at 20 m elevation. Note that the intermediate MIS 3 level terraces (40,000–60,000 year BP) can only be observed if uplift rates are greater than ~1 mm/year.

Figure 5.18 shows a close-up of the previous figure, focusing on the range of uplift rates that produce a terrace at 20 m elevation with an age between 40,000 and 50,000 year BP. Terraces can be cut at 10 m, 20 m and 50 m for several uplift rates within this range. If the uplift rate is 2.0 or 2.25 mm/year, all three terraces would have formed during the same complicated highstand (between 40,000 and 60,000 year BP) MIS 3. *L. lithophaga* borings can be observed to elevations of ~50 m on the south coast of western Crete. Although these shells are cemented in their boreholes, probably by karstic dripstone, they are still clearly recognisable, suggesting that the uplift has been rapid. It would therefore make sense for all three terraces (10 m, 20 m and 50 m) to have been formed during the last (MIS 3) intermediate highstand. Some of the other flat areas seen in Fig. 5.14 may also have formed during MIS 3.

In Fig. 5.17a and b, the clearest terraces correspond to the MIS 5 highstands (100,000–125,000 years BP). For uplift rates approaching those expected for the west of Crete, based on the dates of the 20 m terraces, these MIS 5 terraces should be hundreds of metres above sea-level. Erosion on the south coast of Crete is rapid due to the high uplift rate and steep topographic gradients, so these high terraces are unlikely to be preserved.

If we use the uplift pattern from the AD 365 earthquake, we can estimate the elevation at which terraces should occur on the north coast of Crete. In the earthquake, the south-west corner of the island was uplifted to a maximum elevation of 10 m above sea-level, whereas the north coast was uplifted to just 1–2 m above sea-level. If this co-seismic uplift can be extrapolated to long-term uplift rate, the uplift rate on the north coast should be 1/10 to 1/5 of that on the south coast, i.e, 0.2–0.4 mm/year. At this rate, the 125 ka terrace is the youngest terrace expected to be observed (See Fig. 5.17). The 125 ka terrace should be between 25 m and 50 m

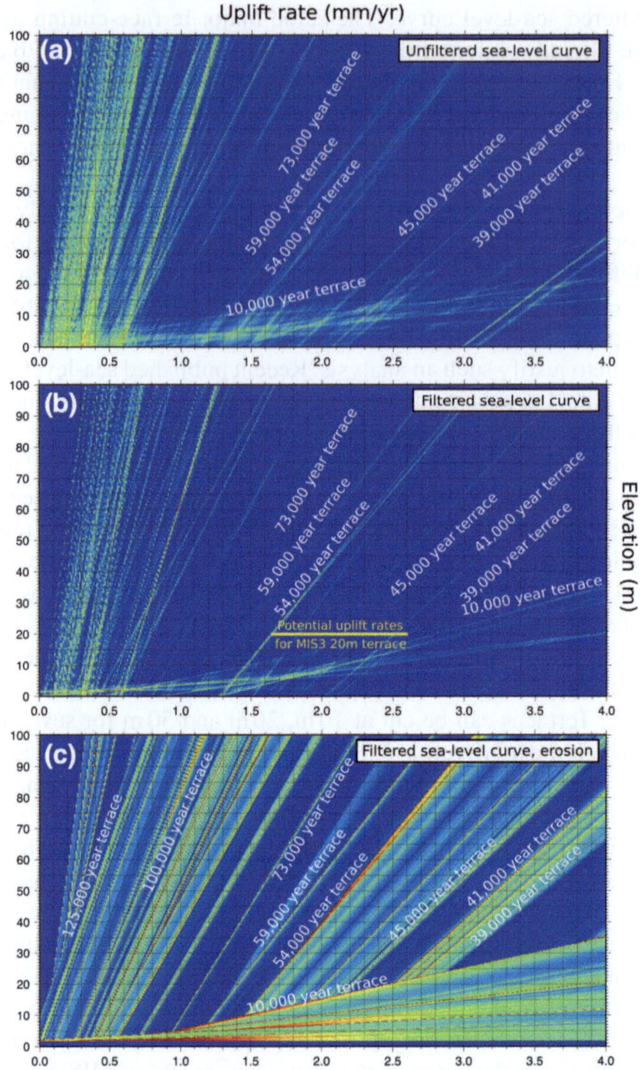

Fig. 5.17 Expected terrace elevation for a range of uplift rates. *Yellow/green lines* mark elevations at which terraces are expected. **a** This plot has been constructed from the unfiltered sea-level curve. **b** The same major terrace-cutting events can be seen when the filtered version of the sea-level curve is used and the data is far less noisy. The *horizontal yellow line* shows the potential uplift rates based on the ages of the 20 and 22 m terraces at Paleochora and Chrisoskalitissa. **c** In this case, destruction of terraces occurs when they are submerged beneath sea-level, so only the youngest terraces are preserved at a given elevation. This plot is ideal for investigating young terrace-cutting events. The colour palette has been modified into enhance the terraces

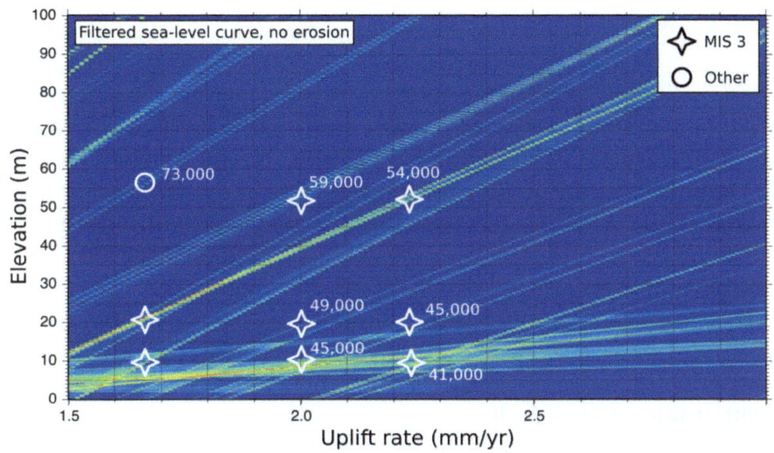

Fig. 5.18 Close-up of the region of interest. Within the range of possible uplift rates, based on the radiocarbon age of the 20–22 m terrace, there are several cases where terraces would be cut at 10, 20 and 50 m. If the uplift rate is ~2 mm/year, the 10, 20 and 50 m terrace could all have been formed during the same complicated highstand corresponding to MIS 3 (between 40,000 and 60,000 years BP)

Fig. 5.19 Possible terrace at 20 m elevation on the north coast. The top of the terrace is at 35° 30.693′N, 23° 57.312′E. **a** Looking E and **b** looking W at the same surface. **c** Image taken from Google Earth showing the locations and directions of the photographs (*white arrows*)

above sea-level. A horizontal surface, which appears to be a terrace, can be observed on the north coast close to Chania at ~20 m above sea-level (Fig. 5.19), but further work is needed to establish whether this is the MIS 5 terrace.

5.7 Conclusions

The main focus of this chapter is an attempt to place constraints on the long-term pattern and rate of uplift of western Crete. I attempted to use kinematic GPS techniques to map accurate river profiles in order to invert for uplift history. However this proved to be unfeasible due to poor satellite reception in constricted river gorges. In contrast, kinematic GPS techniques were useful in mapping uplifted marine terraces. Accurate elevation maps of terraces, combined with radiocarbon dating of uplifted marine organisms, allow inferences to be made about uplift rate.

On a large scale, it is plausible that the tilting and extent of western Crete appears to be controlled by a splay fault surfacing at the Hellenic Trench. However the expression of uplift in the topography is strongly modified by the geology, and by north-south striking normal faults.

Radiocarbon dates of terraces in the south-west corner of Crete at elevations near 20 m suggest that the local uplift rate is ~2 mm/year, and that terraces at similar elevation were formed during marine isotope stage 3. Using the sea-level curve of Rohling et al. [11], I modelled terrace formation for varying uplift rates. An uplift rate close to 2 mm/year results in terraces between 10 m and 60 m that are also likely to have formed during MIS 3. This inference is consistent with the observation that *L. lithophaga* shells are found in situ to elevations of 50 m.

Based on the relative uplift of the north and south coasts during the AD 365 earthquake and an uplift rate on the south-west corner of Crete of 2 mm/year, the uplift rate of the north coast should be between 0.2 and 0.4 mm/year. If that is so, then on the NW coast of Crete the MIS 5 terraces should be observed at elevations between 20 and 40 m above present-day sea-level.

References

1. W. Nemec, G. Postma, Quaternary alluvial fans in southwestern Crete: sedimentation processes and geomorphic evolution. Alluvial Sediment. Special Publ. Int. Assoc. Sediment. **17**, 235–276 (1993)
2. R. Pope, K. Wilkinson, E. Skourtsos, M. Triantaphyllou, G. Ferrier, Clarifying stages of alluvial fan evolution along the Sfakian piedmont, southern Crete: New evidence from analysis of post-incisive soils and OSL dating. Geomorphology **94**, 206–225 (2008)
3. X. Le Pichon, J. Angelier, The Aegean Sea, Philosophical Transactions of the Royal Society of London. Ser. A. Math. Sci. **300**, 357–372 (1981)
4. A.C. Mann, Trace element geochemistry of high alumina basalt - andesite - dacite - rhyodacite lavas of the main volcanic series of Santorini volcano, Greece. Contributions Mineral. Petrol. **84**, 43–57 (1983)
5. L. Briqueu, M. Javoy, J.R. Lancelot, M. Tatsumoto, Isotope geochemistry of recent magmatism in the Aegean arc: Sr, Nd, Hf and O isotopic ratios in the lavas of Milos and Santorini. Earth Planet. Sci. Lett. **80**, 41–54 (1986)
6. G. Zellmer, S. Turner, C. Hawkesworth, Timescales of destructive plate margin magmatism: new insights from Santorini, Aegean volcanic arc. Earth Planet. Sci. Lett. **174**, 265–281 (2000)

7. G.G. Roberts, N. White, Estimating uplift rate histories from river profiles using African examples. J. Geophys. Res. **115** (2010). doi:10.1029/2009JB006692.
8. B. Andronopoulos, Geological map of Greece 1:50,000, Institute of Geology and Mineral Exploration (1993)
9. R. Armijo, H. Lyon-Caen, D. Papanastassiou, East-west extension and Holocene normal fault scarps in the Hellenic arc. Geology **20**, 491–494 (1992)
10. R. Armijo, B. Meyer, G.C.P. King, A. Rigo, D. Papanastassiou, Quaternary evolution of the Corinth Rift and its implications for the late Cenozoic evolution of the Aegean. Geophys. J. Int. **126**, 11–53 (2007)
11. E.J. Rohling, K. Grant, M. Bolshaw, A.P. Roberts, M. Siddall, C. Hemleben, M. Kucera, Antarctic temperature and global sea-level closely coupled over the past five glacial cycles. Nat. Geosci. **2**, 500–504 (2009)
12. L.C. McNeill, R.E.L.I. Collier, Uplift and slip rates of the eastern Eliki fault segment, Gulf of Corinth, Greece, inferred from Holocene and Pleistocene terraces. J. Geol. Soc. London **161**, 81–92 (2004)
13. H. Akima, A method of bivariate interpolation and smooth surface fitting for irregularly distributed data points. ACM Trans. Math. Softw. **4**, 148–159 (1978)
14. K.A. Hughen, J. Southon, S. Lehman, C. Bertrand, J. Turnbull, ^{14}C calibration and activity for the past 50,000 years updated from the Cariaco Basin. Quat. Sci. Rev. **25**, 3216–3227 (2006)

Chapter 6
Conclusions

In spite of its relative quiescence in the instrumental record, the Hellenic subduction zone is capable of producing large magnitude earthquakes, and tsunamis that travel throughout the Mediterranean.

In common with previous studies, I confirm that the subduction zone interface dips at a shallow angle, failing in low-angle thrust-faulting earthquakes in the depth range 15–45 km, generally reaching a maximum depth of 20 km in the west and 45 km in the centre of the arc, near Crete. The subduction zone interface is largely uncoupled, accommodating just 10% of the geodetically measured Nubia-Aegean convergence seismically in the last 100 years.

In contrast, large earthquakes can be generated on more steeply-dipping faults branching from the subduction zone interface at depth. These faults are revealed by steeper-dipping reverse faults at shallower depths above the main interface in most parts of the arc. It was one such splay fault that was responsible for the catastrophic AD 365 earthquake, which uplifted western Crete to a maximum of 10 m above sea-level. This fault surfaces close to the Hellenic Trench, a clear bathymetric escarpment that runs the length of the subduction zone. The position of the splay faults may control the presence, extent and tilting of islands in the sedimentary arc, uplifting the overriding Aegean by sediment underplating. The principal large earthquake and tsunami hazard in the region is likely to be from rare AD 365-type events on splay faults, which may occur as often as every 800 years when the entire arc is considered.

The slip-vectors of shallowly-dipping thrust faulting earthquakes along the subduction zone interface diverge around the arc, and this divergence is closely matched by the GPS measurements. The resulting along-arc extension is taken up by normal faults which strike N–S, and segment the uplifting sedimentary arc.

In the west, the subduction zone terminates in a steep bathymetric escarpment known as the Kefalonia Transform Fault. Strike-slip motion occurs along this fault, which marks a change from oceanic subduction in the SE, to continent–continent collision in the NW. Earthquake mechanisms, GPS data and a flexural gravity anomaly show that the origin of the escarpment is as a lateral ramp, formed as the Ionian Islands are emplaced SW onto the Apulian lithosphere. Shear between Apulia and the

B. Shaw, *Active Tectonics of the Hellenic Subduction Zone*, Springer Theses, DOI: 10.1007/978-3-642-20804-1_6,

Aegean is not confined to the KTF, but is is distributed on several parallel strike-slip faults over a distance of ~150 km. One of these was responsible for the destructive earthquake of 8 June 2008 near Patras.

The GPS velocity field also shows a 150–200 km wide band of NE-SW right-lateral shear extending from the NE Aegean to the SW termination of the Hellenic subduction zone. In the NE and SW this motion is achieved by parallel NE-SW strike-slip faults (branches of the North Anatolian fault in the NE, and the KTF and parallel fault strands in the SW). In both the east and west of central Greece, the strike-slip faults terminate in E-W graben that die out towards the centre of Greece. The rapid gradients of extension along-strike of these graben is the structural manifestation of clockwise rotation that accommodates the shear through central Greece. Central Greece thus acts as a relay zone connecting the strike-slip faults to either side.

Extending the pattern of uplift from the AD 365 earthquake along the arc can be used to predict areas where evidence of long-term uplift might be observed. Clear terraces can be recognised in western Crete, and *Lithophaga lithophaga* burrows can be recognised to at least 50 m above sea-level, and rivers have been forced to downcut rapidly, confirming rapid uplift of the western part of Crete. Terraces have been recognised in the Peloponnese, and the island of Kefalonia appears to have clear uplifted terraces, with well-preserved *Cladocora sp.* corals within the marine cap. However there is no clear notch along the coastline of the Peloponnese, suggesting that if AD 365-type earthquakes do occur in this section of the subduction zone, then the previous earthquake probably occurred more than 6,000 years ago.

Radiometric dating of uplifted organisms is invaluable for reconstructing the tectonic history of a deforming region. I have shown that the commonly collected boring bivalve, *Lithophaga lithophaga* can record radiocarbon ages that are up to 2,000 years older than the uplift event which exposed them. I have shown that they can rarely be used to distinguish uplift events, or date them to better than 1,000 years, or even to distinguish whether observed uplift occurred in a single or multiple events. Lithophagid ages are consistent with the hypothesis that the highest uplifted borings and marine notches in the coastal regions of central Greece were formed about 6,000 years BP, when sea-level became relatively stable near its present level, following a rapid rise at the end of the last glacial maximum.

The rapid and sustained uplift Western Crete is probably related to sediment underplating by slip along faults splaying from the subduction zone interface. In other subduction zones around the world, such as Sumatra, islands above subduction zones appear to behave as predicted by an elastic model of the earthquake cycle. During the interseismic period, the islands close to the trench are pulled down due to locking along the subduction zone interface, and the islands further from the trench are uplifted by the flexural bulge. During the earthquake, the interface releases, and islands close to the trench uplift, and those that were uplifted in the flexural bulge subside. Such behaviour is well-documented in Sumatra, where coral atolls and flooded mangroves attest to cyclical uplift and subsidence [1–3]. In this region, very little sediment enters the subduction zone.

An interesting and rewarding area for future study would be to see whether this difference in sustained vs. cyclical uplift is a first order feature of subduction zones, related to the amount of sediment that is subducting. Close to Alaska, a great deal of sediment enters the subduction zone, and Middleton Island was uplifted by up to 10 m by slip along a splay fault during the 1964 great Alaskan earthquake [4], providing an intriguing parallel with the Cretan AD 365 earthquake.

References

1. C. Subarya, M. Chlieh, L. Prawirodirdjo, J.-P. Avouac, Y. Bock, K. Sieh, A.J. Meltzner, D.H. Natawidjaja, R. McCaffret, Plate-boundary deformation associated with the great Sumatra–Andaman earthquake. Nature **440**, 46–51 (2006)
2. J. Zachariasen, K. Sieh, F.W. Taylor, R.L. Edwards, W.S. Hantoro, Submergence and uplift associated with the giant 1933 Sumatran subduction earthquake: evidence from coral microatolls. J. Geophys. Res. **104**, 895–919 (1999)
3. J. Zachariasen, K. Sieh, F.W. Taylor, W.S. Hantoro, Modern vertical deformation above the Sumatran subduction zone: Paleogeodetic insights from coral microatolls. Bull. Seismol. Soc. Am. **90**, 897–913 (2000)
4. G. Plafker, Tectonic deformation associated with the 1964 Alaska earthquake. Science **148**, 1675–1687 (1965)

Chapter 7
Earthquake Tables

This chapter contains the focal parameters and locations of the earthquakes described and plotted in Chap. 3. They are separated into their tectonic setting, as in Chap. 3. The table references are listed at the end of this appendix.

7.1 Earthquakes in the Downgoing Nubian Plate

7.1.1 Good Waveform-Modelled Earthquakes

yyyy	mm	dd	hh	mm	Lat	Long	Depth	Mw	Strike	Dip	Rake	References
1965	3	31	9	46	38.38	22.26	55	6.8	136	76	78	[6]
1965	4	9	23	57	35.02	24.30	51	6.1	63	76	157	[4]
1969	6	12	15	13	34.39	25.02	19	6.0	163	50	44	[4]
1972	9	13	4	13	37.96	22.33	80	6.1	133	65	164	[6]
1975	9	22	0	44	35.26	26.26	64	5.6	209	75	131	[4]
1979	08	22	20	13	35.910	27.415	68	5.3	64	31	−106	[13]
1982	8	17	22	22	33.77	22.93	39	6.3	230	45	109	[4]
1983	3	19	21	41	35.05	25.29	67	5.6	44	51	139	[4]
1985	9	27	16	39	34.27	26.58	38	5.8	125	77	9	[4]
1987	05	29	18	41	37.541	21.648	49	5.2	51	59	−175	[13]
1992	11	21	05	07	35.93	22.40	54	5.9	194	53	28	[8]
1993	3	18	15	47	38.33	22.09	56	5.7	139	63	61	[8]
1994	5	23	06	46	35.58	24.72	75	6.7	64	77	163	[8]
1996	04	26	07	01	36.34	27.96	70	5.3	343	54	174	[13]
2001	06	23	06	53	35.648	28.102	43	5.6	345	79	177	[10]
2002	01	22	04	54	35.624	26.645	93	6.1	9	36	−176	[13]
2002	05	21	20	53	36.621	24.342	105	5.8	352	89	4	[13]
2006	01	08	11	35	36.31	23.2	60	6.5	195	42	53	[10]
2008	01	06	05	14	36.97	22.87	68	6.0	218	45	20	[10]
2008	07	15	23	52	35.67	27.75	48	6.2	359	65	175	[10]

B. Shaw, *Active Tectonics of the Hellenic Subduction Zone*, Springer Theses,
DOI: 10.1007/978-3-642-20804-1_7, © Springer-Verlag Berlin Heidelberg 2012

7.1.2 Other Waveform-Modelled Earthquakes

yyyy	mm	dd	hh	mm	Lat	Long	Depth	Mw	Strike	Dip	Rake	References
1977	11	28	02	59	36.00	27.80	66	5.6	88	51	32	[13]
1984	05	22	13	57	35.90	22.60	63	5.1	188	44	32	[13]
1987	06	19	18	46	36.80	28.20	65	5.2	121	41	88	[13]
1991	10	18	14	05	35.70	28.56	43	5.2	82	88	13	[13]
1996	04	12	15	39	36.59	27.04	162	5.2	235	80	−52	[13]
1997	07	27	10	08	35.51	21.18	16	5.3	203	86	47	[12]
2002	06	06	22	36	35.54	26.01	94	4.9	153	31	−22	[13]

7.1.3 CMT Solutions

yyyy	mm	dd	hh	mm	Lat	Long	Depth	Mw	Strike	Dip	Rake	References
1983	01	03	00	12	34.547	24.363	74	5.1	30	36	70	[15]
1983	09	27	23	59	36.746	26.919	155	5.4	312	46	162	[15]
1985	07	22	21	32	34.490	28.309	49	5.2	67	48	−34	[15]
1986	10	02	10	12	34.795	28.355	38	5.2	99	37	−53	[15]
1987	04	12	02	47	35.440	23.495	48	5.1	252	90	180	[15]
1988	11	20	21	01	35.379	28.715	31	5.4	24	32	−152	[15]
1997	12	02	19	23	35.958	19.723	18	5.2	206	49	−20	[15]
1998	03	09	11	21	35.982	28.410	60	5.1	10	33	−116	[15]
1999	06	11	07	50	37.645	21.298	49	5.2	304	82	−177	[15]
2000	06	15	21	30	34.484	20.051	34	5.1	42	25	91	[15]
2003	04	29	01	51	36.822	21.805	45	5.1	193	81	4	[15]
2003	09	13	13	46	36.658	26.885	147	5.2	97	40	10	[15]
2004	10	07	01	05	36.553	26.804	130	5.5	334	40	166	[15]
2004	11	04	06	22	35.894	23.216	71	5.2	187	48	58	[15]
2007	02	03	13	43	35.787	22.498	59	5.4	93	68	173	[15]

7.2 Aegean-Nubia Convergence

7.2.1 Good Waveform-Modelled Earthquakes

yyyy	mm	dd	hh	mm	Lat	Long	Depth	Mw	Strike	Dip	Rake	References
1966	5	9	0	42	34.42	26.48	16	5.5	132	46	110	[4]
1972	5	4	21	39	35.17	23.54	41	6.2	112	74	98	[4]
1977	8	18	9	27	35.24	23.49	38	5.5	114	79	96	[4]
1977	9	11	23	19	34.92	23.04	19	5.9	276	47	89	[4]
1979	5	15	6	59	34.54	24.41	35	5.7	253	17	65	[4]
1979	6	15	11	34	34.90	24.14	40	5.6	150	75	70	[4]
1984	6	21	10	43	35.38	23.28	39	6.0	110	72	83	[4]
1986	05	22	19	52	34.607	26.572	27	5.3	118	86	99	[13]
1995	12	07	18	01	34.767	24.123	19	5.1	344	16	126	[10]
1995	12	10	03	28	34.842	24.156	21	5.1	344	16	126	[10]
1997	10	13	13	40	36.374	22.161	20	6.3	300	18	86	[10]
1997	11	05	12	23	34.760	23.962	19	5.1	358	26	153	[10]
1999	04	17	08	17	36.057	21.778	27	5.3	172	59	95	[13]
2000	05	24	05	41	35.998	22.023	18	5.5	111	80	63	[13]
2005	11	25	09	31	35.153	23.451	35	5.1	300	23	72	[10]
2008	02	14	10	09	36.24	21.79	29	6.7	296	8	82	[10]
2008	02	14	12	09	36.29	21.67	32	6.1	302	12	83	[10]
2008	03	28	00	16	34.38	25.31	43	5.4	247	24	61	[10]
2008	06	21	11	36	36.09	21.87	9	5.5	146	61	128	[10]

7.2.2 First Motion Solutions

yyyy	mm	dd	hh	mm	Lat	Long	Depth	Ms	Strike	Dip	Rake	References
1968	05	30	17	40	35.49	27.96	21	6.0	314	25	119	[2]
1969	01	14	23	12	36.18	29.20	33	6.3	100	74	82	[2]
1969	04	16	23	21	35.34	27.77	45	5.5	104	80	85	[2]

7.3 Overriding Aegean

7.3.1 Good Waveform-Modelled Earthquakes

yyyy	mm	dd	hh	mm	Lat	Long	Depth	Mw	Strike	Dip	Rake	References
1965	4	27	14	9	35.62	23.50	13	5.4	191	65	−79	[5]
1973	11	29	10	57	35.14	23.81	18	5.8	224	67	10	[4]
1986	9	13	17	24	37.05	22.12	8	5.8	196	51	−90	[5]
1996	07	20	00	01	36.200	27.069	16	5.9	205	43	−75	[10]
1999	10	05	00	53	36.748	28.218	21	5.0	240	41	−57	[13]
2000	06	13	01	43	35.208	27.054	16	5.0	49	71	−46	[13]

7.3.2 Other Waveform-Modelled Earthquakes

yyyy	mm	dd	hh	mm	Lat	Long	Depth	Mw	Strike	Dip	Rake	References
1992	04	30	11	44	35.10	26.60	7	5.8	214	52	−47	[12]

7.4 Eastern End

7.4.1 Good Waveform-Modelled Earthquakes

yyyy	mm	dd	hh	mm	Lat	Long	Depth	Mw	Strike	Dip	Rake	References
1966	05	09	00	42	34.42	26.48	16	5.5	284	47	70	[4]
1986	05	22	19	52	34.607	26.572	27	5.3	297	10	90	[9]
2000	06	13	01	43	35.208	27.054	16	5.3	157	47	−153	[9]
2006	08	13	10	35	34.46	26.58	17	5.1	197	43	−1	[10]

7.4.2 Other Waveform-Modelled Earthquakes

yyyy	mm	dd	hh	mm	Lat	Long	Depth	Mw	Strike	Dip	Rake	References
1990	07	09	11	22	34.90	26.60	9	5.2	217	56	−21	[12]
1991	03	19	12	09	34.80	26.30	12	5.5	261	30	40	[12]
1992	04	30	11	44	35.10	26.60	7	5.8	214	52	−47	[12]

7.4.3 CMT Solutions

yyyy	mm	dd	hh	mm	Lat	Long	Depth	Mw	Strike	Dip	Rake	References
1988	09	05	20	03	26.599	34.516	15	5.2	15	55	−11	[15]
1989	08	27	01	21	26.277	34.793	45	5.6	223	19	33	[15]
2001	05	29	04	44	27.764	35.477	24	5.1	42	66	13	[15]
2002	10	12	05	59	26.359	34.862	36	5.4	250	29	2	[15]
2005	08	04	10	45	26.497	34.902	10	5.0	90	80	7	[15]
2006	08	22	09	23	27.087	35.155	31	5.1	267	88	100	[15]
2007	05	21	16	39	27.805	35.111	10	5.0	243	69	−56	[15]
2007	09	23	00	54	27.092	35.242	24	5.3	234	41	−23	[15]

7.4.4 First Motion Solutions

yyyy	mm	dd	hh	mm	Lat	Long	Depth	Ms	Strike	Dip	Rake	References
1957	04	24	19	10	36.40	28.60	1	6.8	83	63	28	[2]
1957	04	25	02	25	36.50	28.60	1	7.2	58	85	19	[2]
1968	05	30	17	40	35.49	27.96	21	6.0	314	25	119	[2]
1969	04	16	23	21	35.34	27.77	45	5.5	310	11	116	[2]
1971	01	03	23	18	34.90	26.30	2	5.4	335	20	101	[1]

7.5 Strike-Slip Earthquakes in KTZ

7.5.1 Good Waveform-Modelled Earthquakes

yyyy	mm	dd	hh	mm	Lat	Long	Depth	Mw	Strike	Dip	Rake	References
1972	9	17	14	7	38.25	20.21	8	6.2	39	61	−173	[3]
1983	1	17	12	41	38.01	20.23	11	6.8	48	56	167	[3]
1983	3	23	23	51	38.22	20.28	7	6.2	30	70	176	[3]
1988	05	18	05	17	38.351	20.545	22	5.4	45	70	163	[11]
1989	08	20	18	33	37.338	21.314	16	5.7	193	74	−174	[12]
1992	01	23	04	24	38.372	20.525	9	5.4	18	36	145	[10]
1993	07	14	12	32	38.212	21.826	20	5.4	228	82	−178	[10]
1997	11	18	13	07	37.57	20.66	35	6.2	112	84	43	[8]
2002	12	02	04	59	37.83	21.12	17	5.5	50	68	−166	[14]
2003	08	14	05	15	38.946	20.624	13	6.2	16	72	178	[9]
2007	03	25	13	58	38.36	20.24	11	5.7	31	88	−158	[10]
2008	02	20	18	27	36.24	21.73	12	6.0	254	88	−1	[10]
2008	06	08	12	26	37.97	21.60	20	6.2	295	85	6	[10]

7.5.2 Other Waveform-Modelled Earthquakes

yyyy	mm	dd	hh	mm	Lat	Long	Depth	Mw	Strike	Dip	Rake	References
1981	06	24	18	41	37.87	20.10	20	5.2	27	60	171	[13]
1983	01	19	00	02	38.15	20.22	9.1	5.8	41	49	171	[11]
1983	01	31	15	27	38.18	20.39	12	5.6	41	82	−177	[11]
1983	03	24	04	17	38.10	20.29	18	5.5	62	70	172	[11]
1983	05	14	23	13	38.44	20.33	13	5.6	36	86	167	[11]
1985	05	23	16	02	36.52	22.15	18	5.1	76	73	−33	[13]
1985	09	07	10	21	37.50	21.20	29	5.3	24	57	168	[12]
1987	02	27	23	34	38.42	20.36	13	5.8	26	61	168	[11]
1987	06	10	14	50	37.17	21.39	27	5.4	25	67	176	[12]
1994	02	25	02	30	38.76	20.54	9	5.5	22	58	168	[11]

7.5.3 CMT Solutions

yyyy	mm	dd	hh	mm	Lat	Long	Depth	Mw	Strike	Dip	Rake	References
1981	06	28	17	20	37.831	20.193	22	5.4	201	51	−160	[15]
1988	10	16	12	34	37.877	20.986	16	5.8	301	76	−3	[15]
1989	06	07	19	46	37.991	21.682	24	5.2	154	64	−26	[15]
1994	11	29	14	30	38.742	20.584	22	5.1	185	90	−180	[15]
2002	07	28	17	16	37.940	20.760	16	5.3	7	42	−178	[15]
2002	12	09	09	35	37.839	20.044	21	5.2	255	28	−20	[15]
2003	11	16	07	23	38.252	20.357	10	5.1	266	24	3	[15]
2006	06	21	15	55	38.996	20.608	10	5.0	11	67	−179	[15]
2007	10	27	05	29	37.703	21.409	10	5.1	297	70	−10	[15]

7.5.4 First Motion Solutions

yyyy	mm	dd	hh	mm	Lat	Long	Depth	Ms	Strike	Dip	Rake	References
1965	04	05	03	12	37.70	21.80	34	6.1	226	58	−161	[2]

7.6 Thrust Earthquakes in KTZ

7.6.1 Good Waveform-Modelled Earthquakes

yyyy	mm	dd	hh	mm	Lat	Long	Depth	Mw	Strike	Dip	Rake	References
1959	11	15	17	08	37.78	20.55	13	6.7	134	7	−90	[3]
1963	12	16	13	47	37.10	20.90	6	5.5	291	7	74	[3]
1966	10	29	2	39	38.86	21.05	15	5.5	324	40	48	[3]
1969	7	8	8	9	37.55	20.32	10	5.8	346	13	108	[3]
1973	11	4	15	52	38.86	20.50	23	5.8	324	50	81	[3]
1976	5	11	16	59	37.38	20.32	13	6.4	323	13	90	[3]
1993	03	05	06	55	37.173	21.574	20	5.1	128	59	57	[13]
1994	4	16	23	09	37.44	20.58	15	5.5	268	12	49	[8]
1997	10	13	13	40	36.374	22.161	26	6.3	300	18	86	[10]
1999	04	17	08	17	36.057	21.778	27	5.3	172	59	95	[13]
2000	05	24	05	41	35.998	22.023	18	5.5	111	80	63	[13]
2003	08	14	16	18	38.758	20.610	10	5.2	326	23	56	[10]
2005	01	31	01	05	37.505	20.193	16	6.5	34	14	145	[10]
2005	10	18	15	26	37.622	20.940	17	5.5	15	24	114	[10]
2006	04	11	17	29	37.60	20.83	14	5.3	30	21	122	[10]
2006	04	12	16	52	37.63	20.74	19	5.5	15	23	115	[10]
2008	02	14	10	09	36.24	21.79	29	6.7	296	8	82	[10]
2008	02	14	12	09	36.29	21.67	32	6.1	302	12	83	[10]
2008	06	21	11	36	36.09	21.87	9	5.5	146	61	128	[10]

7.6.2 Other Waveform-Modelled Earthquakes

yyyy	mm	dd	hh	mm	Lat	Long	Depth	Mw	Strike	Dip	Rake	References
1983	07	14	02	54	35.67	21.81	17	5.3	318	48	124	[12]
1985	04	21	08	50	35.70	22.20	25	5.2	149	50	78	[13]
1989	08	24	02	13	37.94	20.14	16	5.3	36	46	142	[11]
1994	01	11	07	23	35.83	21.83	14	5.3	331	60	126	[12]
1996	02	01	17	57	37.76	19.86	19	5.6	173	55	71	[11]
1998	04	29	03	31	35.96	21.88	13	5.3	176	81	117	[12]
1998	05	01	04	00	37.62	20.75	13	5.1	19	53	131	[12]
1998	10	06	12	28	37.13	20.98	9	5.2	308	61	54	[12]

7.6.3 CMT Solutions

yyyy	mm	dd	hh	mm	Lat	Long	Depth	Mw	Strike	Dip	Rake	References
1986	06	08	04	55	36.025	21.570	22	5.1	109	34	86	[15]
2006	04	03	00	50	37.570	21.008	20	5.0	18	39	126	[15]
2006	04	04	22	05	37.625	21.022	17	5.5	357	26	109	[15]
2006	04	19	15	16	37.659	20.921	19	5.4	345	33	126	[15]
2006	05	25	23	14	36.878	20.452	35	5.2	346	23	129	[15]
2008	02	26	10	47	36.10	21.86	12	5.4	286	34	74	[15]
2008	02	26	16	11	36.07	21.80	18	5.0	321	35	114	[15]
2008	03	23	20	11	36.21	21.73	12	5.0	279	31	69	[15]

7.6.4 First Motion Solutions

yyyy	mm	dd	hh	mm	Lat	Long	Depth	Ms	Strike	Dip	Rake	References
1953	08	12	09	23	38.30	20.80	10	6.9	163	34	101	[7]
1968	03	28	07	39	37.80	20.90	6	5.9	120	71	65	[7]
1973	01	05	05	49	35.80	21.90	40	5.6	136	60	94	[2]
1976	06	12	00	59	37.50	20.60	8	5.8	115	70	90	[7]

7.7 Unclassified

yyyy	mm	dd	hh	mm	Lat	Long	Depth	Mw	Strike	Dip	Rake	References
1982	06	22	03	04	37.163	21.364	30	5.5	16	57	−62	[13]
2000	02	22	11	55	34.579	25.541	29	5.1	12	73	178	[10]
2004	03	17	05	21	34.584	23.412	4	5.9	75	63	164	[10]

References

1. T. Taymaz, J. Jackson, D. McKenzie, Active tectonics of the north and central Aegean sea. Geophys. J. Int. **106**, 433–490 (1991)
2. D. McKenzie, Active tectonics of the Mediterranean region. Geophys. J. R. Astron. Soc. **30**, 109–185 (1972)
3. C. Baker, D. Hatzfeld, H. Lyon-Caen, E. Papadimitriou, A. Rigo, Earthquake mechanisms of the Adriatic sea and western Greece: implications for the oceanic subduction-continental collision transition. Geophys. J. Int. **131**, 559–594 (1997)
4. T. Taymaz, J. Jackson, R. Westaway, Earthquake mechanisms in the Hellenic Trench near Crete. J. Int. **102**, 695–731 (1990)
5. H. Lyon-Caen, R. Gaulon, P. Papadimitriou, R. Armijo, J. Drakopoulos, N. Delibassis, V. Kouskouna, K. Makropoulos, J. Baskoutass, J. Latoussakis, D. Papanastassiou, G. Pedotti, The 1986 Kalamata (South Peleponnesus) earthquake: detailed study of a normal fault, evidence for east–west extension in the Hellenic arc. J. Geophys. Res. **93**, 14967–15000 (1988)
6. Y. Liotier, Modelisation des ondes de volume des seismes se L'Arc Egeen. (DEA de l'Universite Joseph Fourier, Grenoble, France, 1989)
7. H. Anderson, J. Jackson, Active tectonics of the Adriatic region. Geophys. J. R. Astron. Soc. **91**, 937–983 (1987)
8. J. Parke, Ph.D. thesis (2001)
9. C. Benetatos, A. Kiratzi, Z. Roumelioti, G. Stavrakakis, G. Drakatos and I. Latoussakis. The 14 August 2003 Lefkada Island (Greece) earthquake: Focal mechanisms of the mainshock and of the aftershock sequence. 2005, 9, 171–190
10. B. Shaw, J. Jackson, Earthquake mechanisms and active tectonics of the Hellenic subduction zone. Geophys. J. Int. (2010). doi: 10.1111/j.1365-246X.2010.04551.x
11. E. Louvari, A.A. Kiratzi, B.C. Papazachos, The Cephalonia transform fault and its extension to western Lefkada Island (Greece). Tectonophysics **308**, 223–236 (1999)
12. A. Kiratzi, E. Louvari, Focal mechanisms of shallow earthquakes in the Aegean sea and the surrounding lands determined by waveform modelling: a new database. J. Geodyn. **36**, 251–274 (2003)
13. C. Benetatos, A. Kiratzi, C. Papazachos, G. Karakaisis, Focal mechanisms of shallow and intermediate depth earthquakes along the Hellenic arc. J. Geodyn. **37**, 253–296 (2004)
14. Z. Roumelioti, Ch. Benetatos, A. Kiratzi, G. Stavrakakis, N. Melis, A study of the 2 December 2002 (Mw 5.5) Vartholomio (western Peleponnese, Greece) earthquake and of its largest aftershocks. Tectonophysics **387**, 65–79 (2004)
15. CMT solution

Chapter 8
Waveform-Modelled Solutions

This chapter contains full solutions for the waveform-modelled earthquakes described in Chap. 3 and listed in the tables in Chap. 7. The title for each figure is the date, given as yymmdd, and the subtitle is strike/dip/rake/depth/M_o (Figs. 8.1, 8.2, 8.3, 8.4, 8.5, 8.6, 8.7, 8.8, 8.9, 8.10, 8.11, 8.12, 8.13, 8.14, 8.15, 8.16, 8.17, 8.18, 8.19, 8.20, 8.21, 8.22, 8.23).

B. Shaw, *Active Tectonics of the Hellenic Subduction Zone*, Springer Theses,
DOI: 10.1007/978-3-642-20804-1_8,

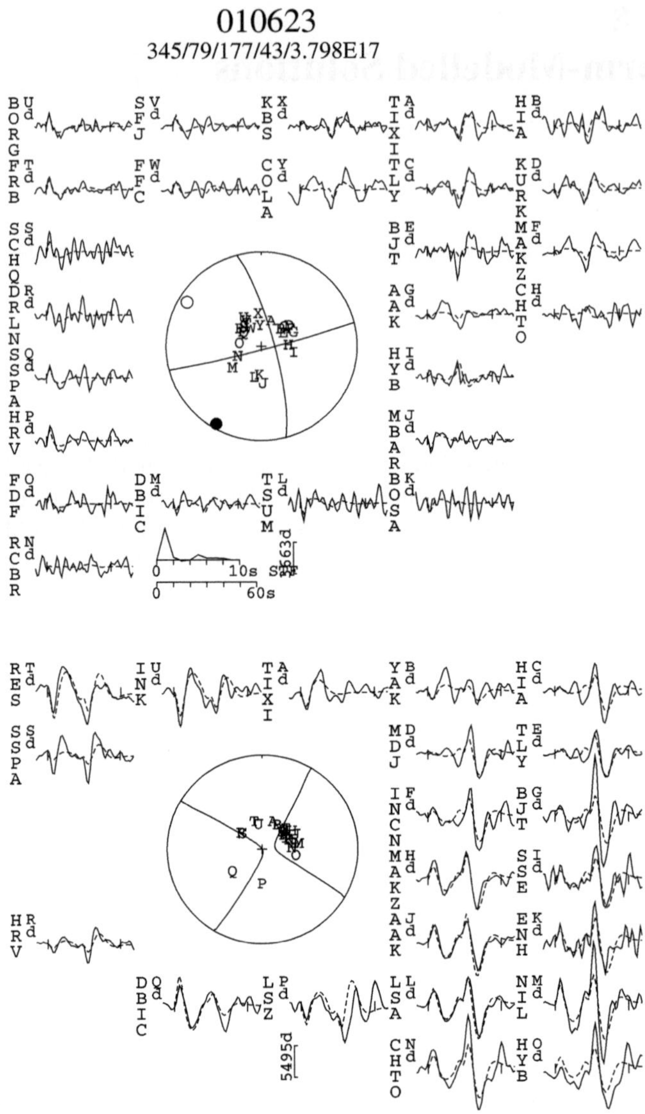

Fig. 8.1 Downgoing Nubian lithosphere: 23 June 2001

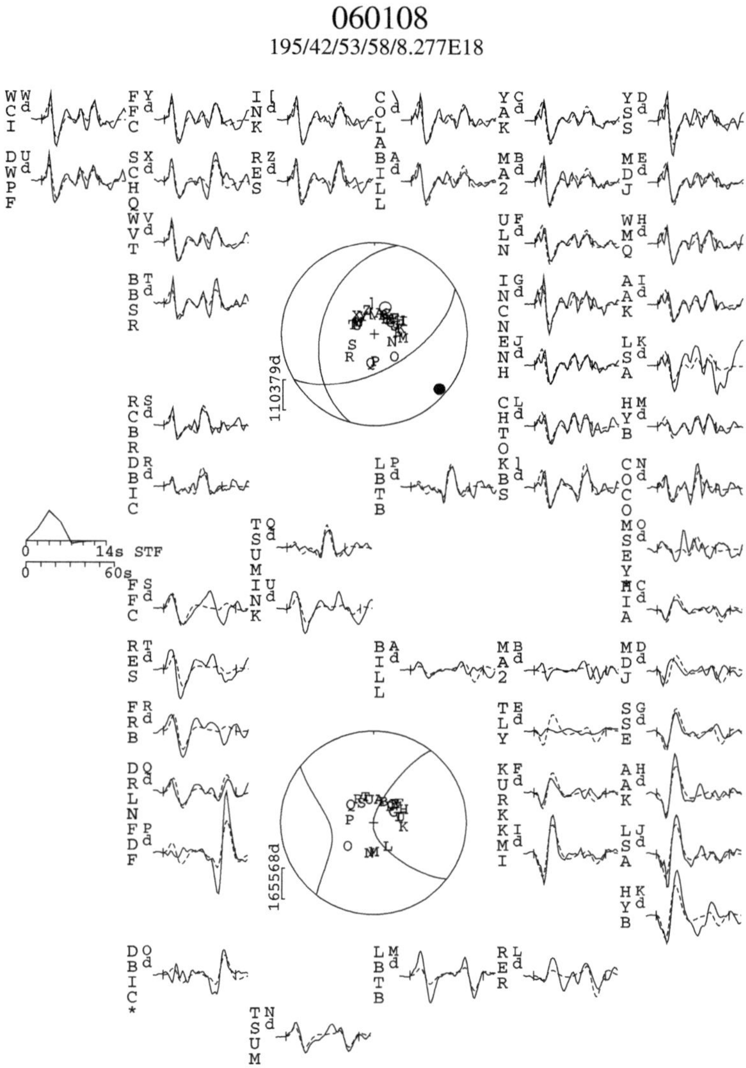

Fig. 8.2 Downgoing Nubian lithosphere: 6 January 2008

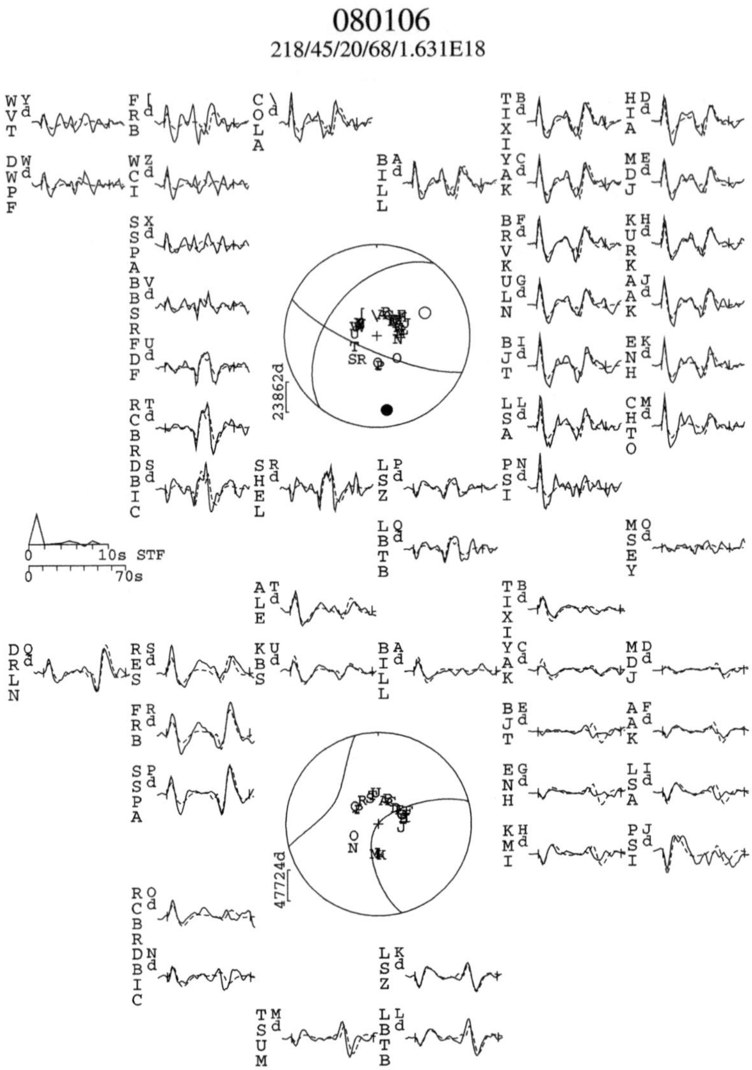

Fig. 8.3 Downgoing Nubian lithosphere: 6 January 2008

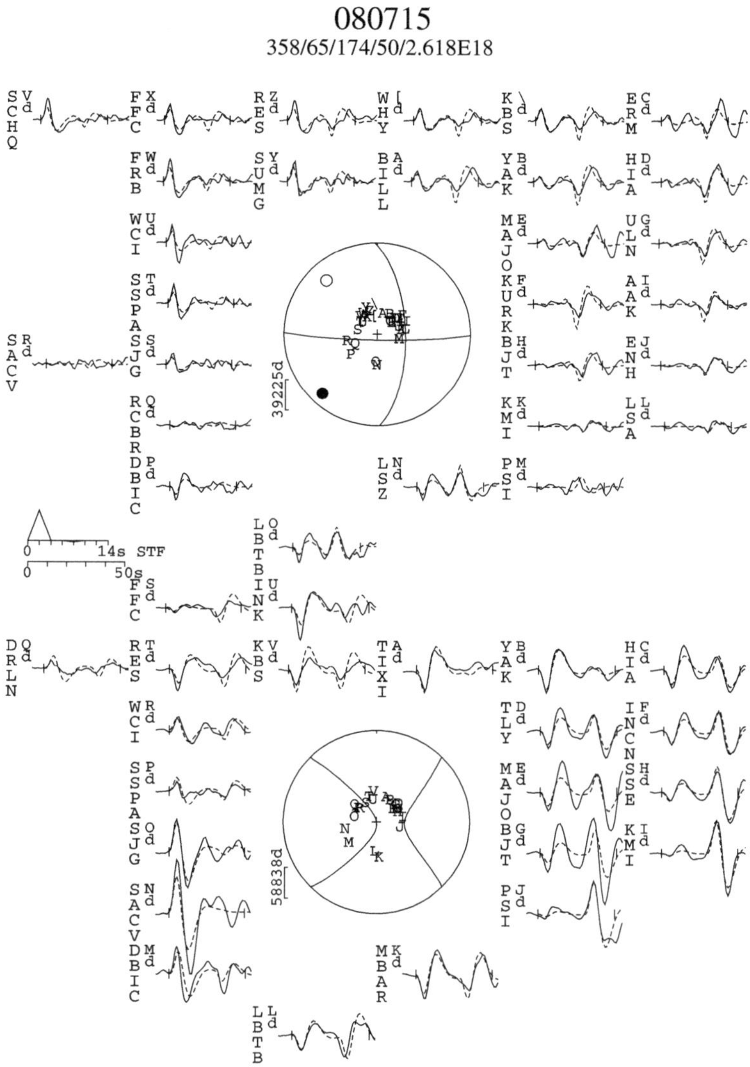

Fig. 8.4 Downgoing Nubian lithosphere: 15 July 2008

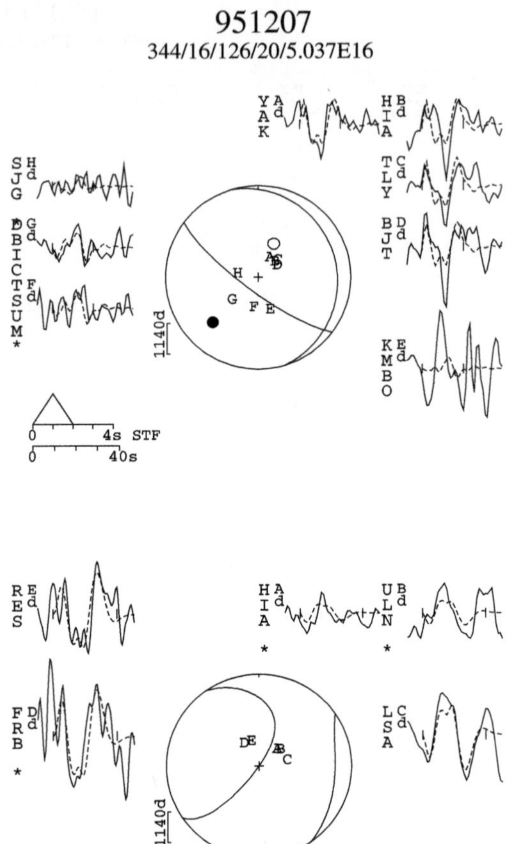

Fig. 8.5 Aegean-Nubian convergence: 7 December 1995 (compare with 10 December 1995)

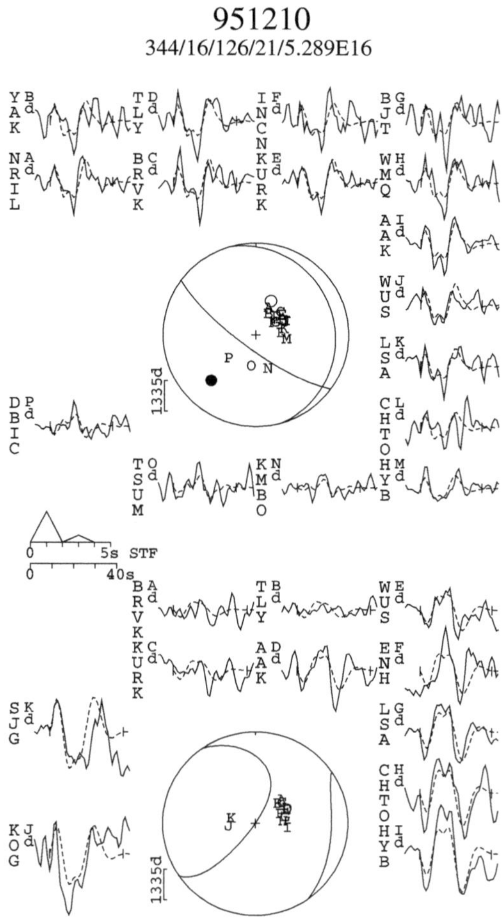

Fig. 8.6 Aegean-Nubian convergence: 10 December 1995

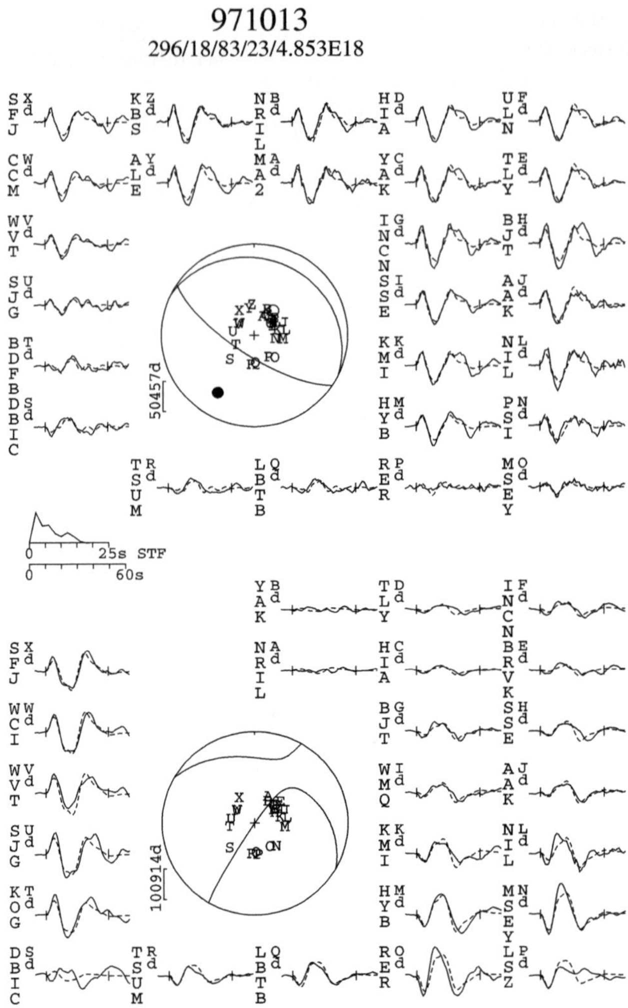

Fig. 8.7 Aegean-Nubian convergence: 13 October 1997

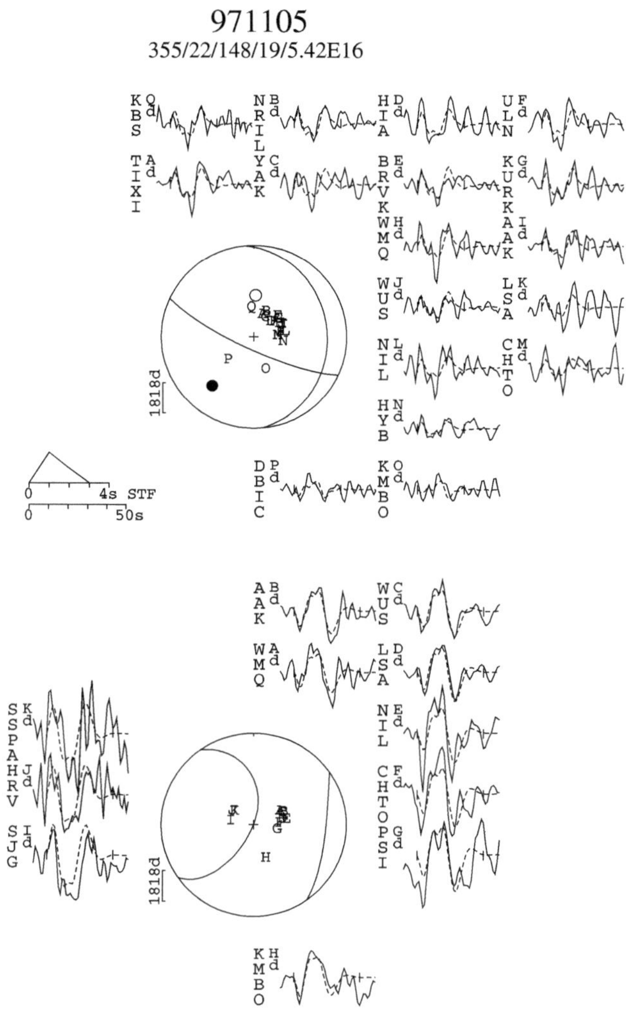

Fig. 8.8 Aegean-Nubian convergence: 13 October 1997

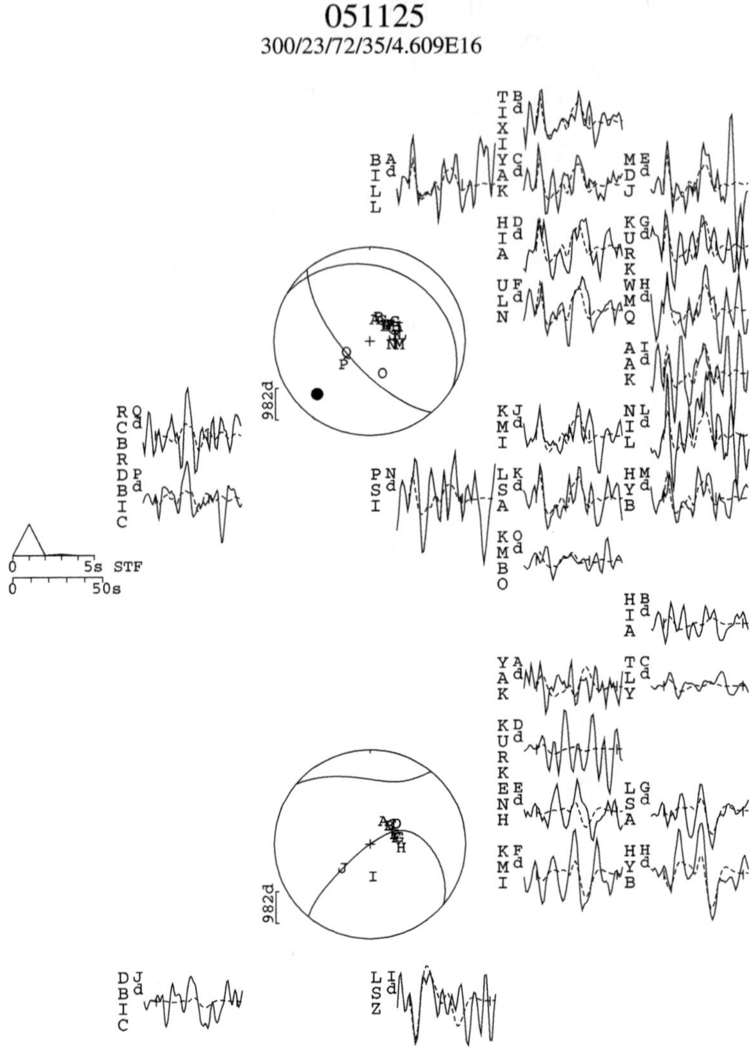

Fig. 8.9 Aegean-Nubian convergence: 25 November 2005

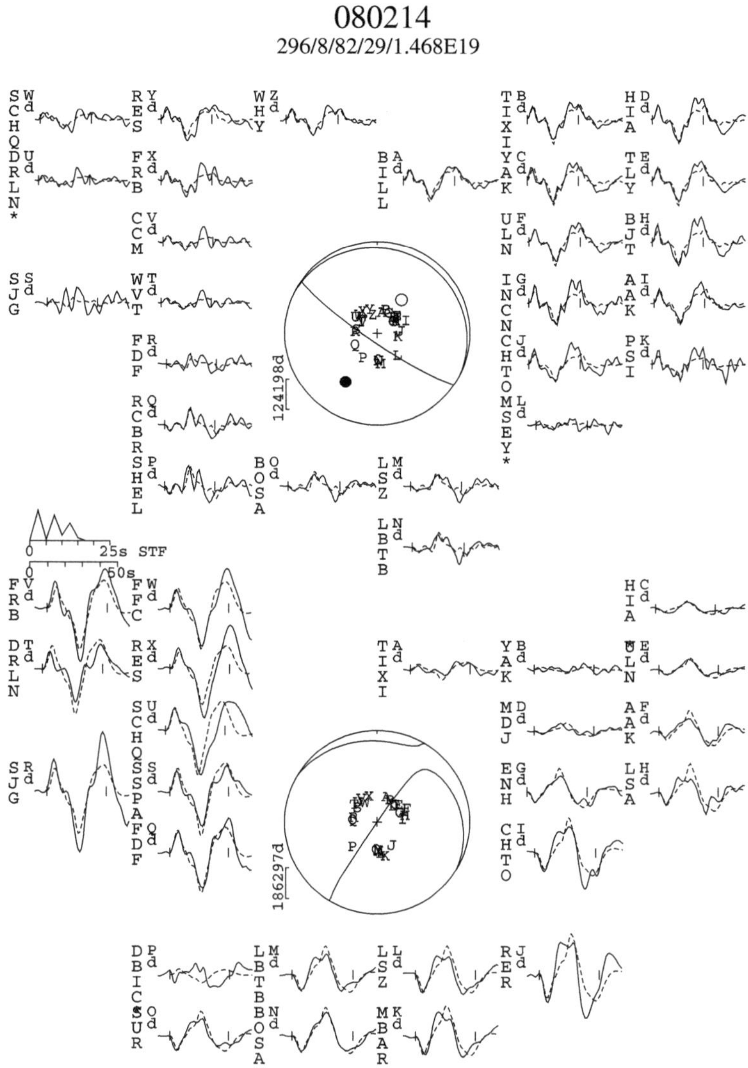

Fig. 8.10 Aegean-Nubian convergence: 14 February 2008

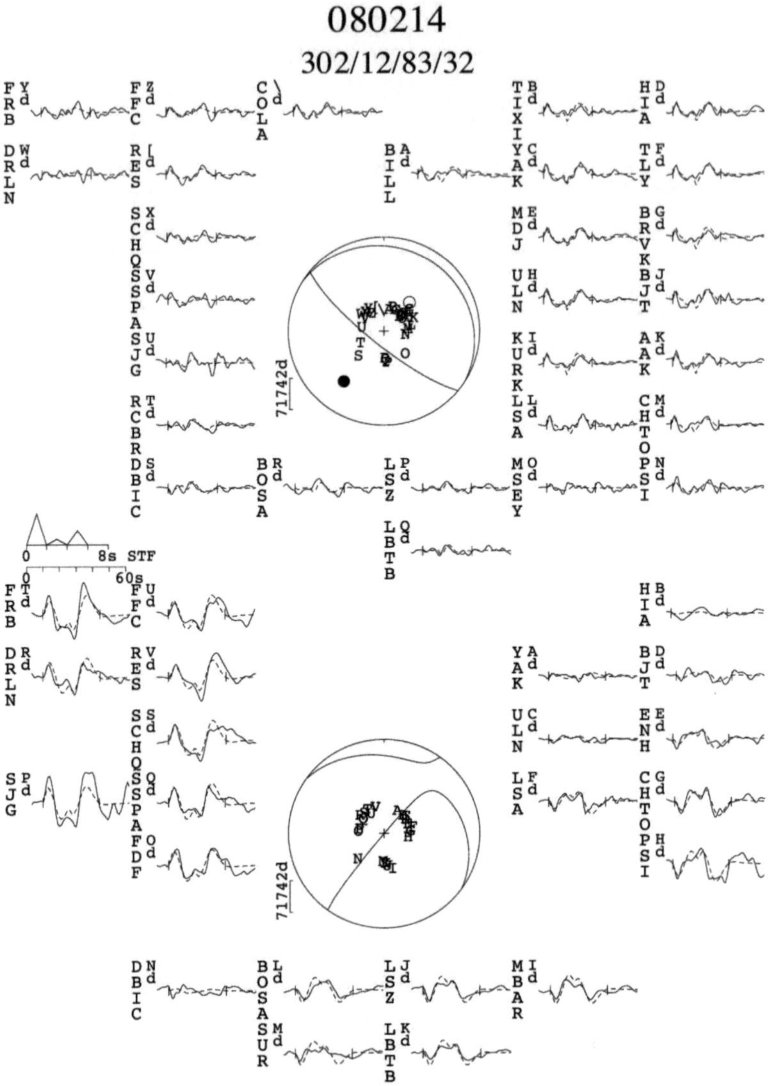

Fig. 8.11 Aegean-Nubian convergence: 14 February 2008

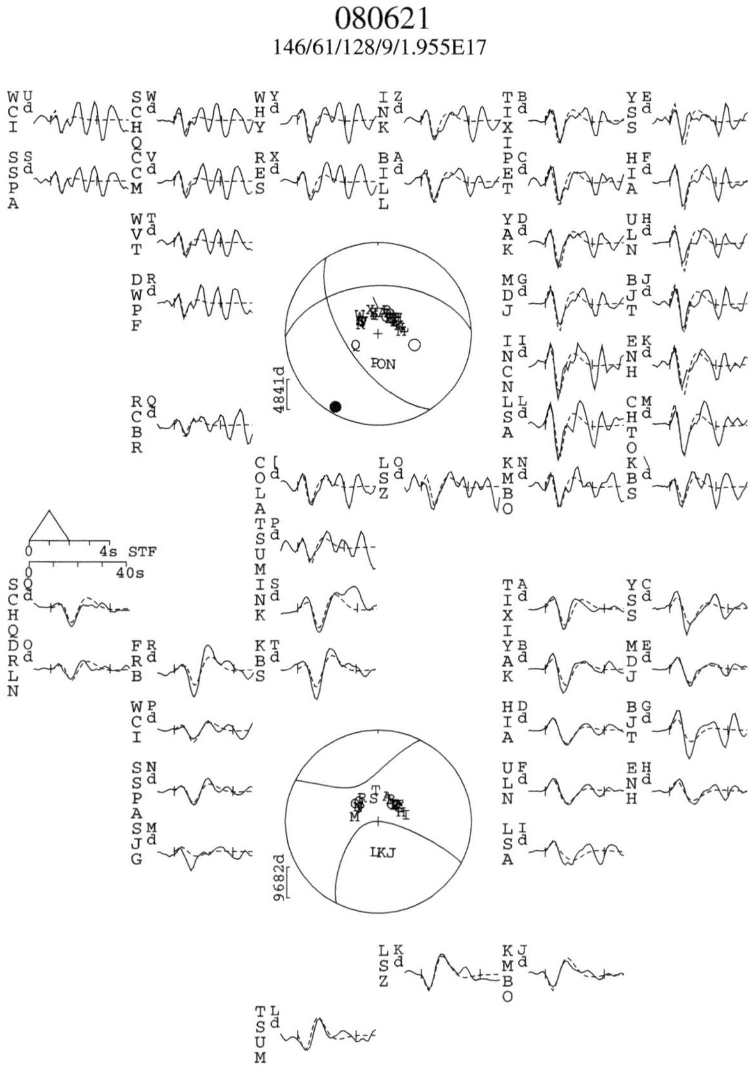

Fig. 8.12 Aegean-Nubian convergence: 21 June 2008

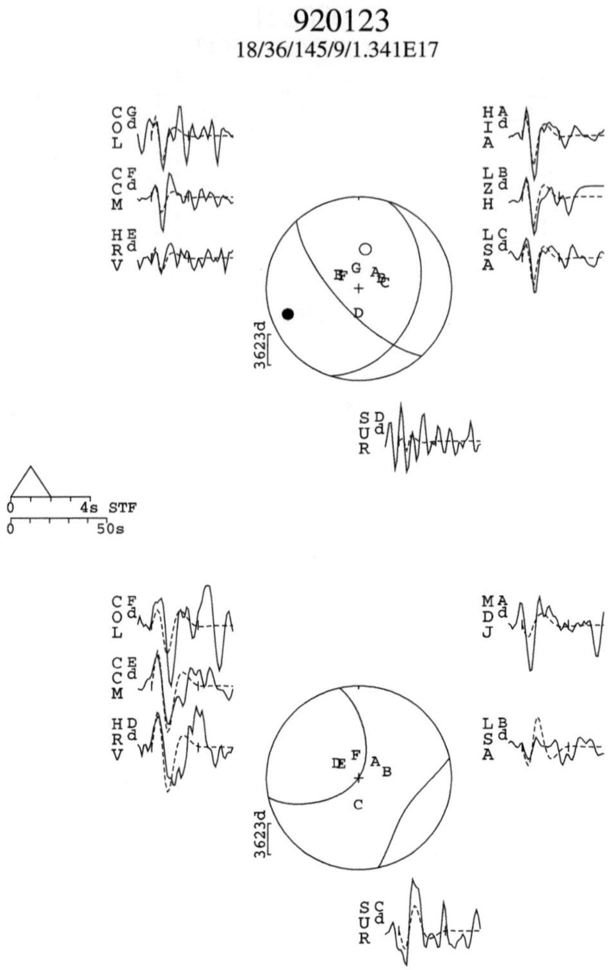

Fig. 8.13 Kefalonia transform region: 23 January 1992

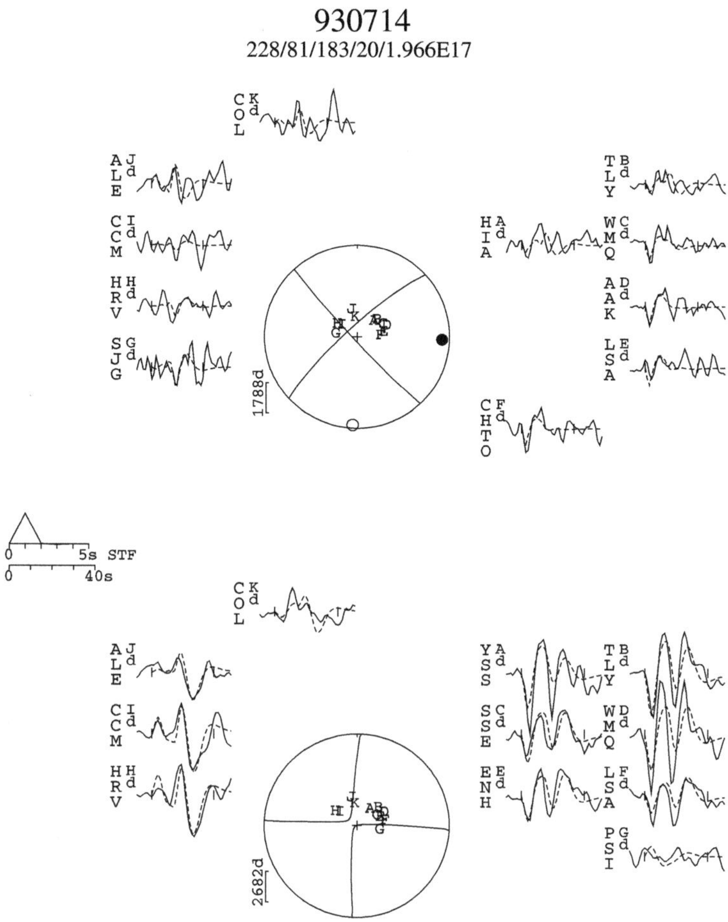

Fig. 8.14 Kefalonia transform region: 14 July 1993

070305

31/88/202/15/4.318E17

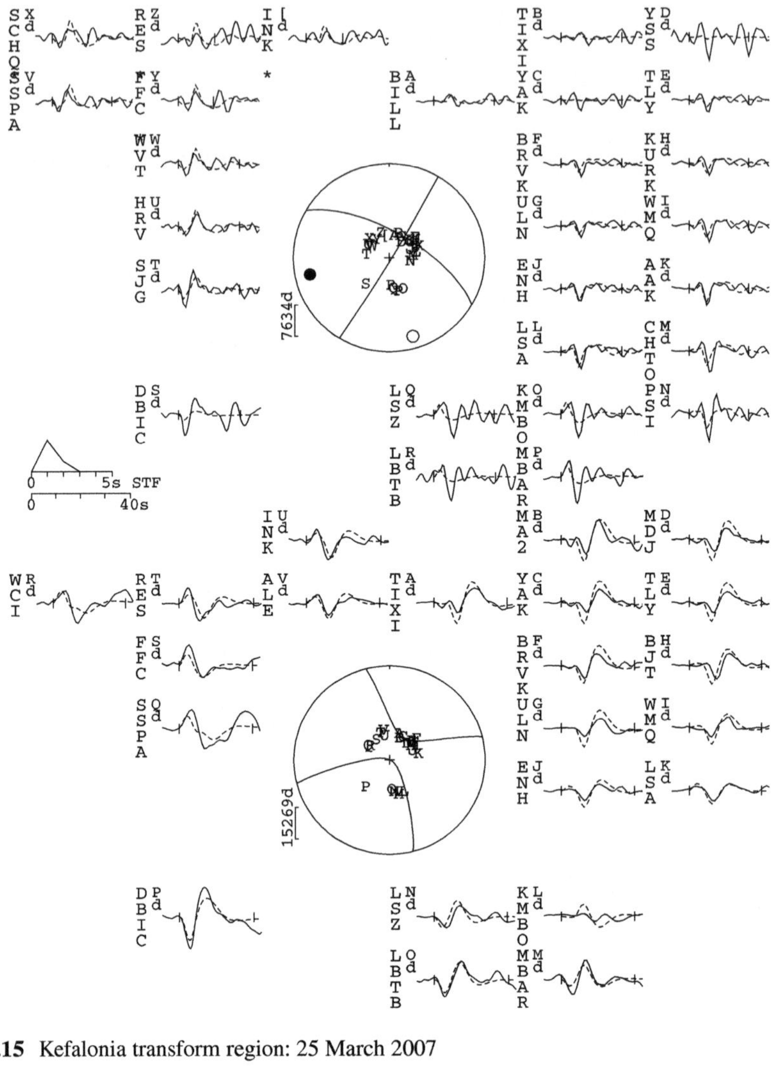

Fig. 8.15 Kefalonia transform region: 25 March 2007

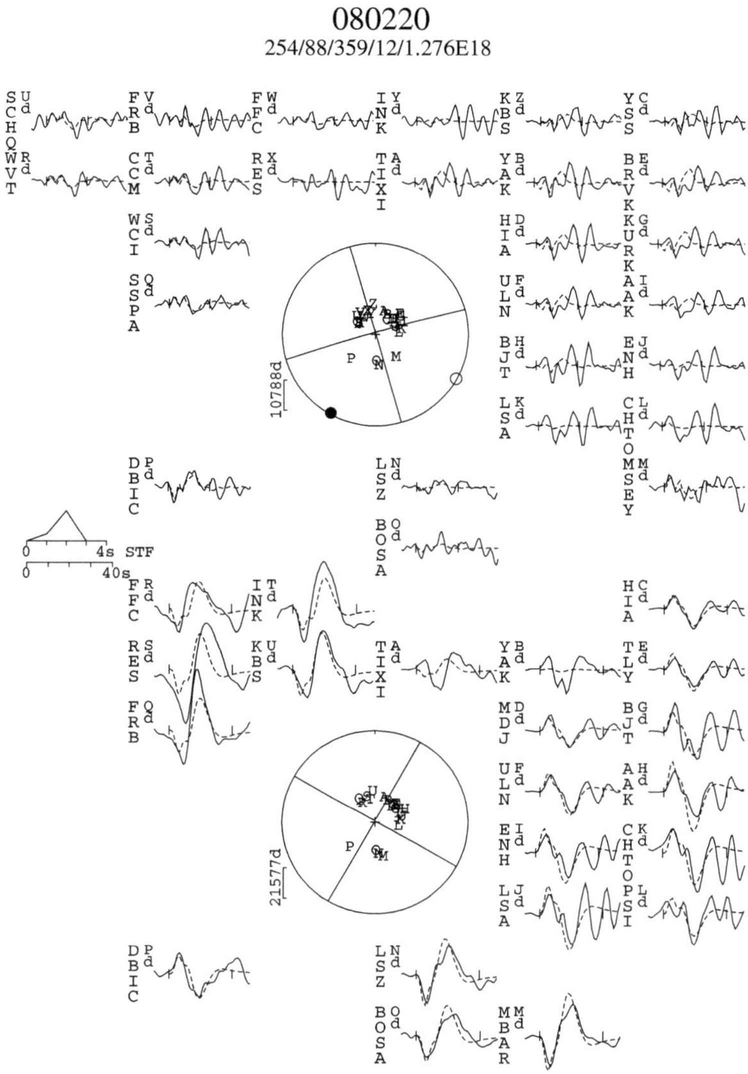

Fig. 8.16 Kefalonia transform region: 20 February 2008

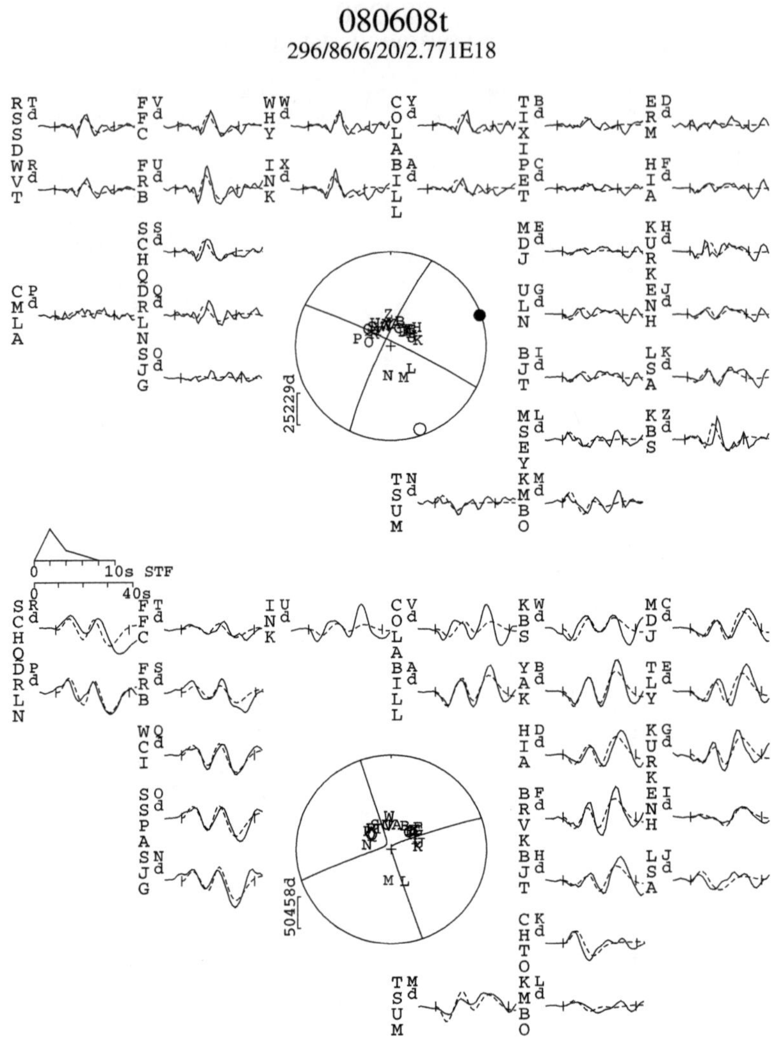

Fig. 8.17 Kefalonia transform region: Patras earthquake, 8 June 2008

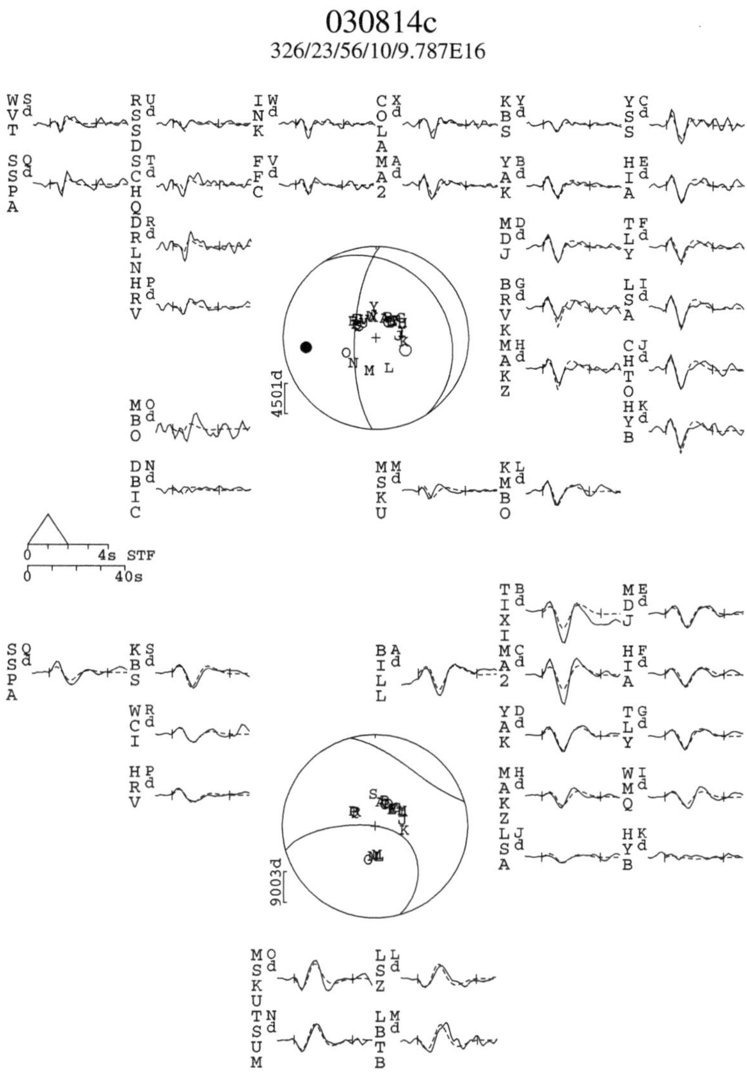

Fig. 8.18 Kefalonia transform region: 14 August 2003

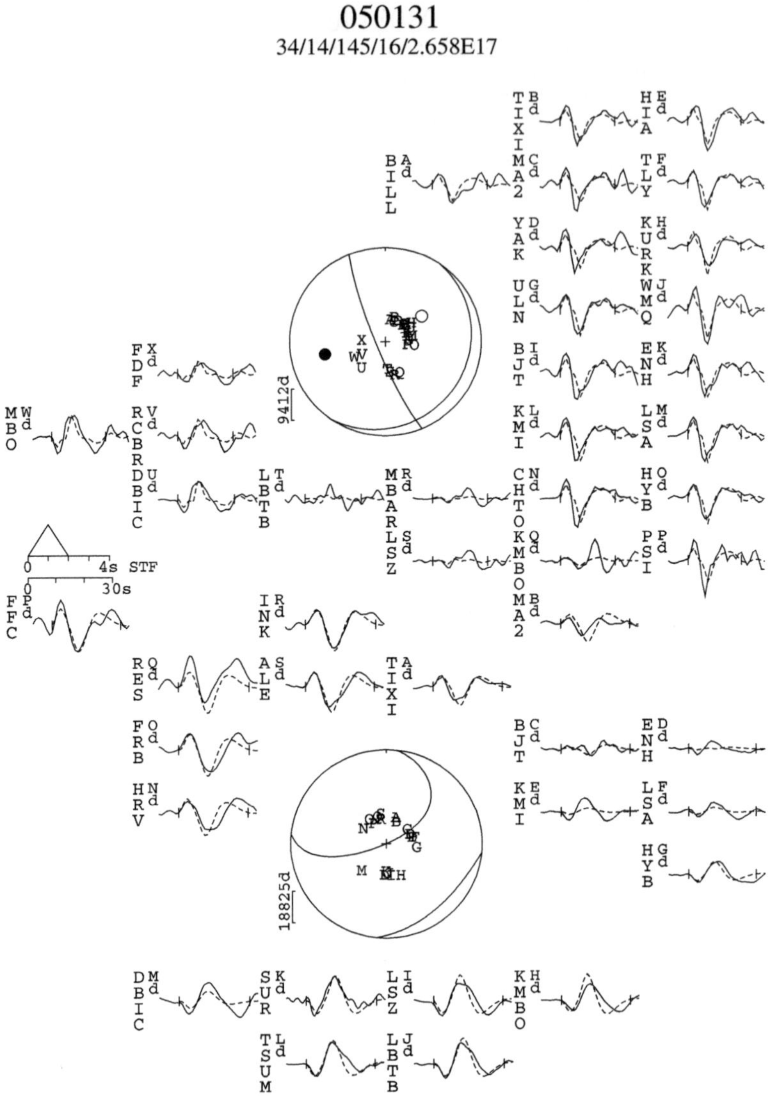

Fig. 8.19 Kefalonia transform region: 31 January 2005

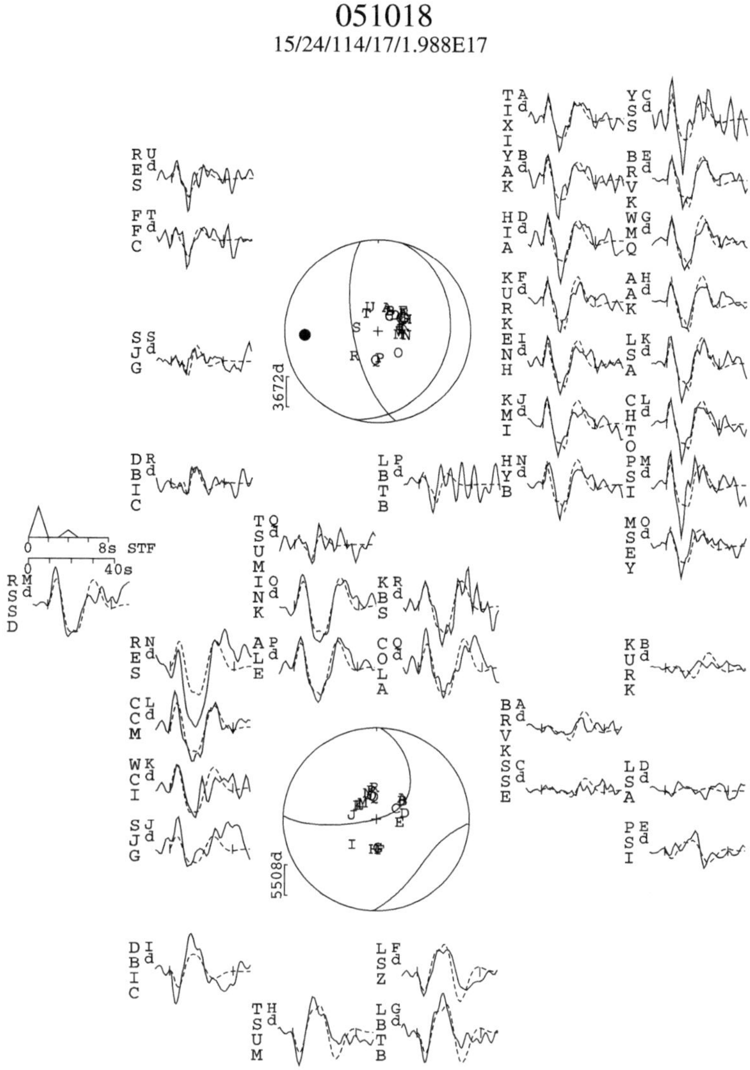

Fig. 8.20 Kefalonia transform region: 18 October 2005

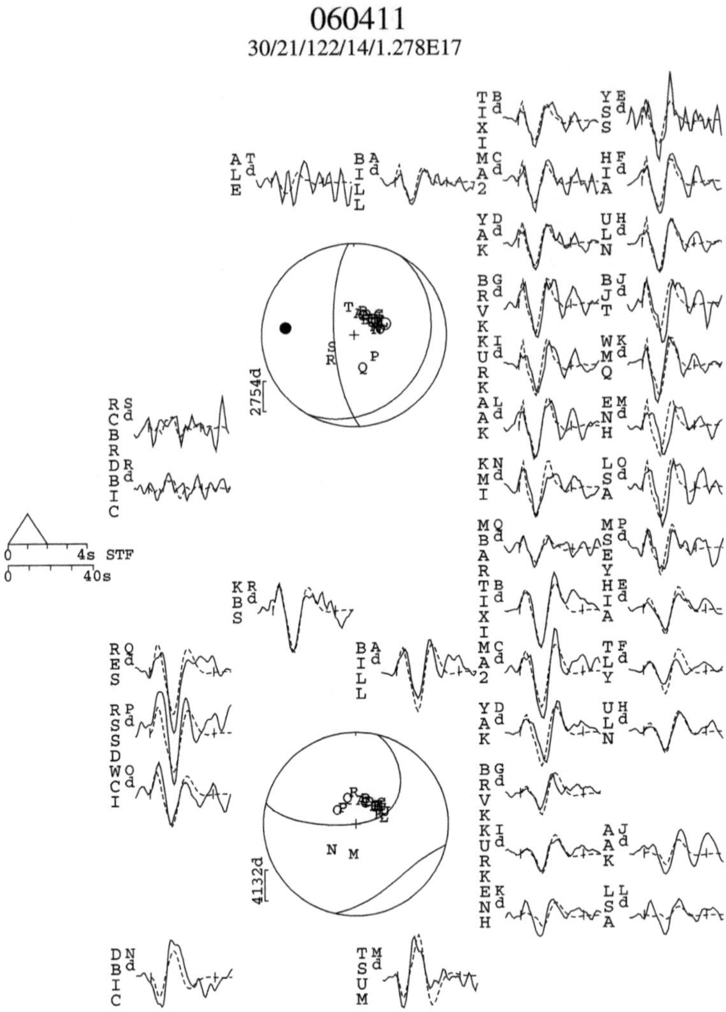

Fig. 8.21 Kefalonia transform region: 11 April 2006

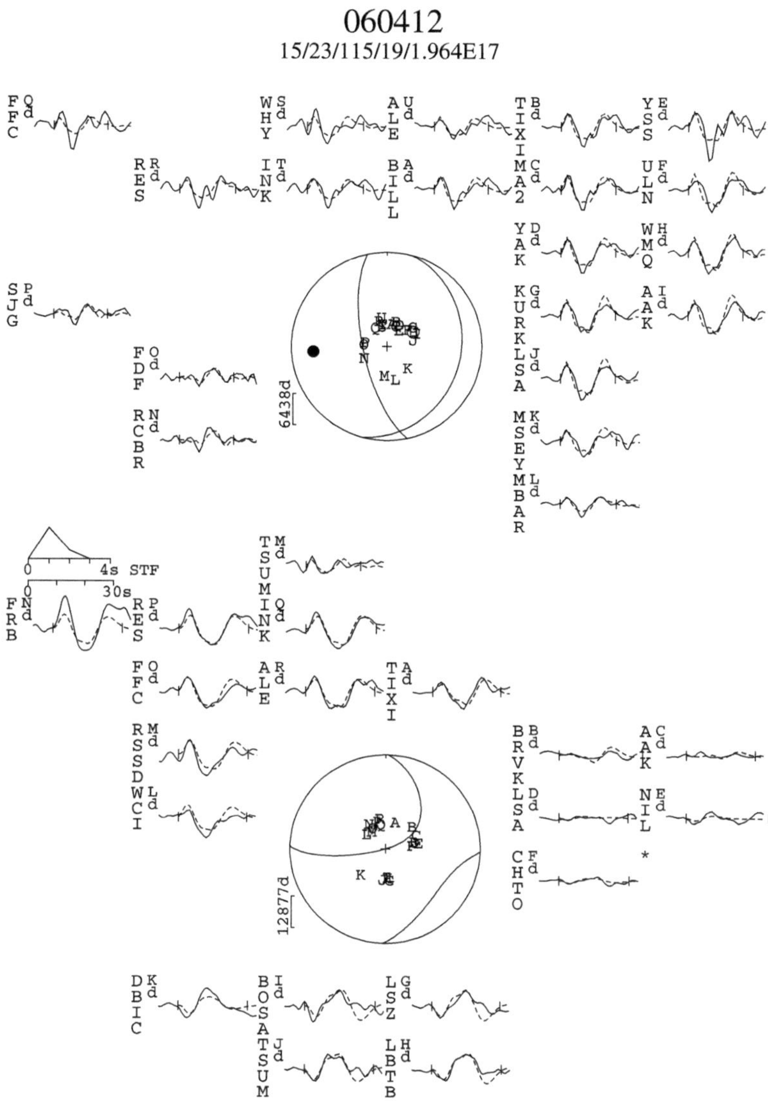

Fig. 8.22 Kefalonia transform region: 12 April 2006

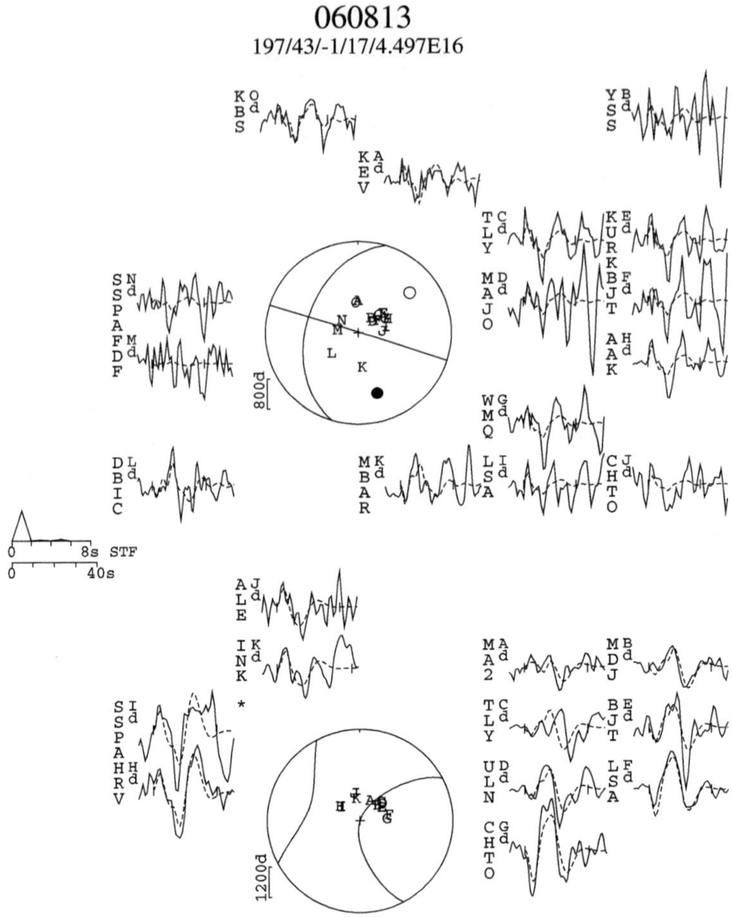

Fig. 8.23 Eastern end: 13 August 2006

Chapter 9
Field Sites

In this chapter, I describe some of the most important field sites that I have visited in Western Crete.

The sites on the north and west coasts are easy to access from Chania using good roads. The ancient harbour at Phalasarna lies at the end of a dirt road through olive and citrus orchards, but is clearly signposted and can be reached without a 4×4. The road along the Grambousa peninsula is very rough and has degraded over the time that I have visited Crete, so is best attempted using a 4×4.

Access to sites on the south coast is possible by road, although there is no road along the south coast itself. To travel from Paleochora to Agh. Roumeli in a single day is possible by boat (ferries run regularly and fishermen are usually happy to give you a lift if you enquire in the harbour at Paleochora), and a spectacular hiking trail runs along the coast from Paleochora to Agh. Roumeli. This hiking trail is invaluable for travelling between Paleochora, Lissos and Soughia, with excellent views of the uplifted shoreline along the way.

I have found that hiring a boat and fisherman for the day should not cost more than 90€ and is an excellent way of exploring the coastline, especially the smaller bays where the regular ferries do not stop (Fig. 9.1).

9.1 Kolymbari to Grambousa

Afrata Bay (35° 34.42′N, 23° 46.61′E) lies just to the north of Kolymbari, on the Rhodopos peninsula. Here the top of the encrustation is at 4.1 m above sea-level. On the west side of the Rhodopos peninsula, ~5 km to the west of Kolymbari, we collected a coral at Agh. Marina Bay, where the AD 365 notch is at an elevation of 5.5 m above sea-level.

AGM	35° 32.11′N	23° 43.53′E	3.0 m	Coral

B. Shaw, *Active Tectonics of the Hellenic Subduction Zone*, Springer Theses, DOI: 10.1007/978-3-642-20804-1_9, © Springer-Verlag Berlin Heidelberg 2012

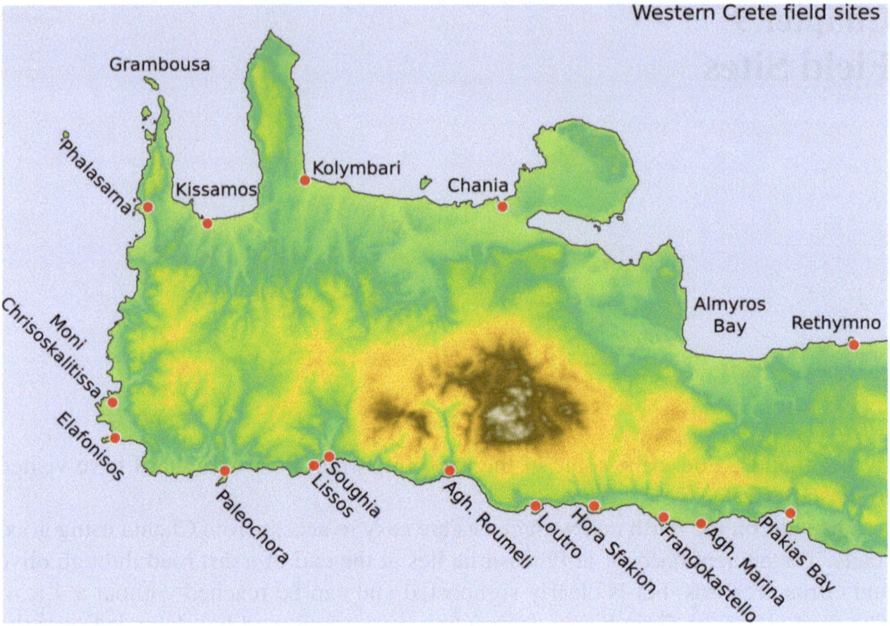

Fig. 9.1 Western Crete field sites and major towns

Fig. 9.2 The AD 365 shoreline on the east side of the Grambousa peninsula (at Agh. Sostis Bay, 35° 35.52′N, 23° 36.10′E) and taken from the western side of the peninsula (35° 35.244′N, 23° 35.682′E), looking towards Imeri Grambousa

9.2 Grambousa Peninsula

Agh. Sostis bay is at the end of the Grambousa peninsula, on the eastern side (35° 35.52′N, 23° 31.10′E). Here the AD 365 shoreline is at an elevation of 6 m above sea-level. On the western side of the peninsula lies Imeri Grambousa island, which is connected to the peninsula by a spit of sand at low tide. The island is topped by a castle, and the AD 365 shoreline at its base was described by Spratt (Fig. 9.2).

9.3 Phalasarna

Phalasarna is the site of an ancient harbour (35° 30.593′N, 23° 34.115′E) that has been extensively excavated. The stone rings, thought to have been used to tie up boats, are heavily encrusted by serpulids, giving an accurate past sea-level. The harbour has been infilled by a fine-grained sediment, which contains multiple horizons of gastropod and bivalve shells (some fragmented, some entire shells).

It is possible to reach the present-day beach by walking to the west of the harbour. Here the AD 365 shoreline is well-preserved at 7 m above sea-level, with a thick band of algal encrustation. There are many overhangs and cave-like features between the harbour and the present-day shore that would have been below sea-level prior to AD 365. Corals were found within *Lithophaga* holes in the protected environment of these overhangs, and *Lithophaga* shells were also found. Radiocarbon dating was carried out on the following samples:

PHA 1	35° 30.574′N	23° 34.097′E	3.5 m	Coral
PHA 2	35° 30.574′N	23° 34.097′E	3.4 m	Coral
PHA 3	35° 30.574′N	23° 34.097′E	3.0 m	Coral
PHA 2	35° 30.574′N	23° 34.097′E	3.4 m	*Lithophaga*

Between Phalasarna and Moni Chrisoskalitissa is Sfinari, where I sampled a *Lithophaga* beneath the AD 365 shoreline:

SFI 1	35° 24.923′N	23° 33.622′E	2 m	*Lithophaga*

9.4 Moni Chrisoskalitissa

The monastery at Chrisoskalitissa has an excellent view of the 22 m terrace from the top of the building.

0.5 km to the north of the monastery lies an almost circular bay (35° 18.900′N, 23° 32.040′E), which appears to have formed as a karstic sinkhole. To the south of the bay (looking towards the monastery—see photograph), there is an extensive exposure of the AD 365 shoreline with a thick algal encrustation. Above the bay lies the 22 m terrace, with a trig point on top, which I mapped using kinematic GPS. A bivalve sample was collected from the marine cap of the terrace, and produced a radiocarbon date of ~40,000 years BP (Chap. 5, Table 5.1) (Fig. 9.3).

Bryozoan, coral and *Lithophaga* samples were collected here from between 2.5 m and 6.0 m above sea-level (CHS 1–CHS 7). Of these samples, four were radiocarbon dated:

Fig. 9.3 Location and photograph of Chrisoskalitissa. Taken from 35° 18.900′N, 23° 32.040′E, looking south

CHS 1	35° 18.857′N	23° 32.008′E	3.3 m	Bryozoan
CHS 2	35° 18.857′N	23° 32.008′E	3.0 m	Bryozoan
CHS 5	35° 18.845′N	23° 31.990′E	2.5 m	Bryozoan
CHS 7	35° 18.857′N	23° 32.115′E	6.0 m	Coral

9.5 Elafonisos

Evidence of the AD 365 uplift can be observed at an elevation of $\sim$ 9 m above the beach at Elafonisos. There are also suggestions of lower notches.

9.6 Paleochora

Evidence of recent, rapid uplift of the west of Crete can be seen at a number of sites close to the town of Paleochora. The town is located on a peninsula, and the main beach is on the west of this peninsula. Beachrock with what appears to be marine cement (isopachous) was collected from this beach (Fig. 9.4).

The 20 m terrace above the harbour is described in Chap. 5, Table 5.1. A bivalve collected from this terrace gave a radiocarbon date of 50,000 years before present. *Lithophaga* bore-holes could be observed to an elevation of at least 30 m.

Along the road to the west of the town, past a petrol station, the clear sea-caves and terraces can be seen to the north of the road. The AD 365 shoreline is on the northern side of the road, and samples were collected from beneath the shoreline (to the south of the road) at the following localities:

On Cape Grameno (3 km to the west of Paleochora), I collected a number of Bryozoan samples, two of which were radiocarbon dated:

Fig. 9.4 Beachrock west of Paleochora (35° 13.992′N, 23° 40.632′E)

PAL 1	35° 14.262′N	23° 39.142′E	5.2 m	*Lithophaga*
PAL 3	35° 14.262′N	23° 39.142′E	9.0 m	*Lithophaga*
PAL 5	35° 14.262′N	23° 39.142′E	3.6 m	*Lithophaga*
PAL 6	35° 14.262′N	23° 39.142′E	1.6 m	*Lithophaga*
PAL 7	35° 14.262′N	23° 39.142′E	6.0 m	*Lithophaga*
PAL 8	35° 14.262′N	23° 39.142′E	9.0 m	Coral

GRAM 2	35° 13.735′N	23° 38.165′E	4.7 m	Bryozoan
GRAM 4	35° 13.736′N	23° 38.186′E	2.0 m	Bryozoan

In Koundoura, approximately 5 km to the west of Paleochora travelling along the coastal road, I climbed the headland to an elevation of at least 70 m above sea-level, and saw clear *Lithophaga* borings to heights in excess of 50 m.

I collected two *Lithophaga* samples here, one of which was radiocarbon dated:

KOU 1	35° 14.200′N	23° 35.625′E	2.5 m	*Lithophaga*

9.7 Lissos

The bay at Lissos can be reached either by walking from Paleochora, or from Soughia. The walk from Soughia passes an archaeological site with beautiful mosaics, and ends at the church of Agh. Kirikos. The top of the AD 365 shoreline is very clear here, at 6.5 m above sea-level, and marked by a thin algal encrustation (Fig. 9.5).

I collected and dated two *Lithophaga* shells at Lissos:

Fig. 9.5 AD 365 shoreline at Lissos (35° 14.395′N, 23° 39.989′E)

KIR 1	35° 14.395′N	23° 39.989′E	2.2 m	*Lithophaga*
KIR 2	35° 14.395′N	23° 39.989′E	3.2 m	*Lithophaga*

9.8 Soughia

To the west of Soughia is a small harbour. The AD 365 shoreline can be clearly seen on both sides of the harbour (see photographs) at 6.7 m above sea-level. The shoreline is marked by a thin band of algal encrustation here and at the end of the eastern beach where it is also observed. Both at the harbour, and at the beach to the east of Soughia, algal encrustation similar to that at the top of the AD 365 shoreline can be seen growing today at sea-level.

Inside the harbour, presumably protected from erosion by waves and storms, a number of horizons can be seen beneath the top of the AD 365 shoreline. These horizons may represent the vertical limits of various species of destructive organism, most of which have a limited vertical range.

A small cave within the harbour contained abundant *Lithophaga* shells, and a colonial coral on the roof (Fig. 9.6).

Lithophaga and Bryozoan samples were collected and dated here:

SOU 1	~ 35° 14.755′N	~ 23° 49.275′E	6.5 m	*Lithophaga*
SOU 1	~ 35° 14.755′N	~ 23° 49.275′E	2.9 m	*Lithophaga*
SOU NB2	35° 14.864′N	23° 49.078′E	2.2 m	Bryozoan

Fig. 9.6 AD 365 shoreline and location map of Soughia. The harbour is at ~35° 14.755′N, ~23° 49.275′E, the beach at 35° 14.864′N, 23° 49.078′E. The bay is ~1 km long

9.9 Agh. Roumeli

Agh. Roumeli is built on sediments carried out of the Samaria gorge, a slot canyon just 1 m wide at its narrowest point. The AD 365 shoreline can be recognised here, and *Lithophaga* bore-holes continue up the cliff face.

9.10 Hora Sfakion

Two *Lithophaga* were sampled close to Hora Sfakion (at Sweetwater Bay):

HSF 1	35° 12.026′N	24° 06.341′E	1.0 m	*Lithophaga*
HSF 2	35° 12.026′N	24° 06.341′E	2.1 m	*Lithophaga*

9.11 Frangokastello

Uplifted and incised alluvial fans dominate the coastline at Frangokastello. The AD 365 shoreline is difficult to recognise, and the substrate is unsuitable for *Lithophaga* borings.

Fig. 9.7 Fans at Frangokastello. The first photograph taken from the main road to Frangokastello from the north, looking SSE

To the east of Frangokastello, the AD 365 shoreline can again be recognised at low elevations (1–2 m) and is marked by a deep erosive notch rather than an algal encrustation. At Plakias, the notch is at an elevation of 2 m above sea-level and has a distinctive double-notch profile. This notch was followed, at decreasing elevation, to Agh. Pavlos where it was observed at 1.6 m above sea-level. No notch could be found at Agh. Galini, probably due to the coarse conglomeritic bedrock (Fig. 9.7).

9.12 Kefalonia and Zakynthos: Reconnaissance Trip

I undertook a reconnaissance survey of Kefalonia and Zakynthos to see whether the rapid recent uplift that controls the topography of Crete is also occurring in the Ionian islands, which mark the transition from oceanic subduction to continental collision.

In 1953 a M_s 7.2 earthquake uplifted the island of Kefalonia by a maximum of 0.7 m, and devastated most of the settlements on both Kefalonia and Zakynthos. A clear uplifted notch associated with this earthquake can be recognised on the cliff-faces around the island and a higher second Holocene notch can also be seen. Pleistocene terraces vary in elevation between 6.5 and 21.6 m on the south coast have also been described [1–3]. Additionally, geological maps of the region shows a number of raised beaches and other Quaternary features around the coast of the island.

I therefore focused my trip on the areas where quaternary features had been recognised, and sought to determine whether there is evidence for the continual uplift of Kefalonia and Zakynthos.

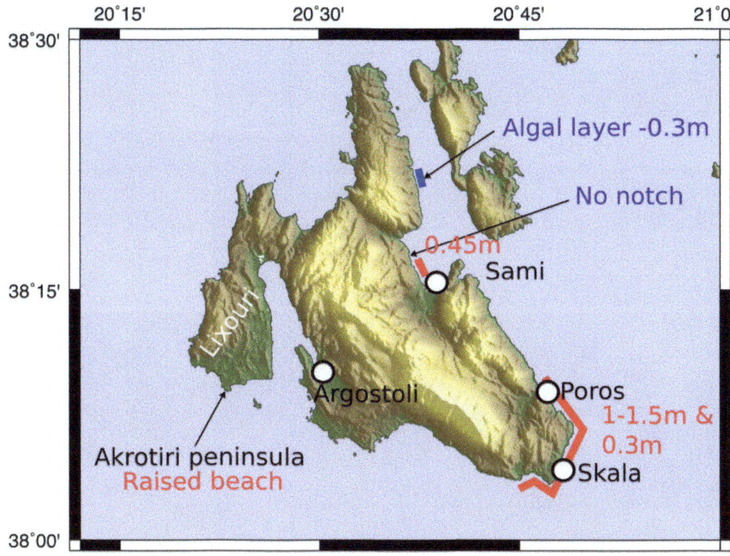

Fig. 9.8 Topographic map of Kefalonia, showing the positions where notches and raised beaches were observed

Fig. 9.9 The first two photographs (*top line*) were taken at Melissani, 38° 15.693′N, 20° 37.465′E and show a very clear notch with *L. lithophaga* holes and encrusting serpulids between the top of the notch and present-day sea-level. The top of the notch is at an elevation of 0.45 m above sea-level. The third photograph is taken at 38° 17.682′N, 20° 36.373′E, further to the north, where no notch was visible. Even further to the north, the purple-pink algae that are generally abundant colonisers of the splash zone formed a clear band 30 cm below sea-level (38° 21.859′N, 20° 36.859). This could possibly be due to extremely high tide when I visited the bay (between 2.00 and 3.15 pm on Saturday 27 September 2008), or could result from recent subsidence by this amount

Fig. 9.10 These photographs were taken between the main harbour at Poros and 38° 09.926′N, 20° 45.940′E. A double erosive notch can be recognised with the higher notch at 1.5 m above sea-level (presumable representing the 6,000 year BP notch) and the lower one between 0.3 above sea-level (revealing the pattern of uplift in the 1953 earthquake). There is also a clear platform in the consolidated limestone at an elevation of 3 m above sea-level whose origin is unclear

Fig. 9.11 Photographs of double erosive notch taken between Poros and Skala. The higher notch is at 1.2–1.5 m above sea-level, the lower at 0.3 m

Fig. 9.12 To the north of Skala, close to 38° 05.97′N, 20° 48.479′N, I observed a flat erosional surface which had been cut into dipping marls, and now lies at 20 m above sea-level. There was no obvious marine cap on top of the surface

Fig. 9.13 The airport is built on a wide, flat plain ∼ 10 m above sea-level. The limestone bedrock outcrops in small patches on the headlands, where it dips ∼ 45° to the SW. Everywhere else, the flat region is underlain by subhorizontal yellow and blue marls. Although the top of the surface is heavily tarmaced and vegetated, fallen blocks on the beach were suggestive of a recent marine cap, containing corals, *L. lithophaga* and shell fragments. The top of the marls certainly looks like an erosional feature, suggesting that the airport is built on a marine terrace

Fig. 9.14 A raised beach at the end of the Akrotiri peninsula. The top of the beach is at an elevation of 5.6 m above sea-level and below this it appears to have been quarried. Yellow marls are overlain unconformably by this beach

9.12.1 Kefalonia

See Figs. 9.8–9.14. If the 1.5 m notch represents 6,000 year BP, the uplift rate here would be 0.25 mm/year, and the elevation of the notch could have been caused by five earthquakes like that in 1953, with a repeat time of 1,200 years. Given this uplift rate, the 125,000 year BP notch (corresponding to the last glacial maximum) should be at an elevation of 31 m above present day sea-level. The hard fossiliferous cap on the yellow/blue marls of the Lixouri Peninsula (Fig. 9.15) is at an elevation of $\sim$30 m above sea-level.

Fig. 9.15 A clear elevated flat area can be seen from the road at 38° 05.97′N, 20° 48.479′E, at a bearing of 134°. Here there is a hard fossiliferous cap at an elevation of 30 m on the yellow/blue marls, and the cap contains serpulids, corals, bivalves and shelly fragments. The corals resemble *Cladocora caespitosa*. The cap therefore seems to represent a raised reefal assemblage of unknown age

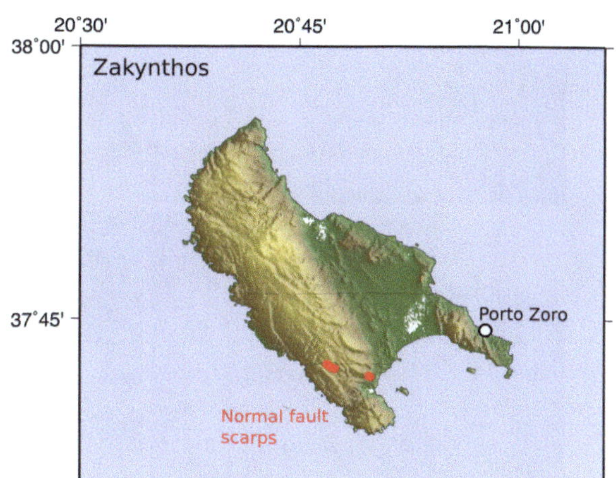

Fig. 9.16 Map of Zakynthos showing the position of Zoro beach, and of the normal fault scarps observed

Fig. 9.17 A very deep notch could be recognised 95 cm above sea-level at Zoro beach, with a lower notch 30 cm above sea-level

Fig. 9.18 Examples of observations of the normal fault scarps. s/d/r = 116/58/76 measured at the site of the first photograph (37° 41.856′N, 20° 49.908′E). See table for GPS points of observations

9.12.2 *Zakynthos*

See Figs. 9.16, 9.17. Normal fault scarps, which appeared to have been formed relatively recently were observed at the following points: Fig. 9.18.

37° 41.856′N	20° 49.908′E
37° 41.862′N	20° 49.896′E
37° 41.886′N	20° 49.854′E
37° 41.904′N	20° 49.776′E
37° 42.300′N	20° 47.472′E
37° 42.300′N	20° 47.454′E
37° 42.318′N	20° 47.418′E
37° 42.324′N	20° 47.394′E
37° 42.336′N	20° 47.352′E
37° 42.336′N	20° 47.346′E
37° 42.426′N	20° 47.160′E
37° 42.444′N	20° 47.130′E
37° 42.456′N	20° 47.082′E
37° 42.480′N	20° 47.058′E
37° 42.486′N	20° 47.046′E
37° 42.504′N	20° 47.028′E
37° 42.516′N	20° 46.998′E
37°42.540 ′N	20° 46.956′E
37° 42.546′N	20° 46.944′E
37° 42.342′N	20° 47.346′E
37° 42.348′N	20° 47.310′E
37° 42.390′N	20° 47.262′E
37° 42.444′N	20° 47.130′E

References

1. P.A. Pirazzoli, S.C. Stiros, J. Laborel, F. Laborel-Deguen, M. Arnold, S. Papageorgiou, C. Morhangel, Late-Holocene shoreline changes related to palaeoseismic events in the Ionian Islands, Greece. The Holocene **4**, 397–405 (1994)
2. P.A. Pirazzoli, J. Laborel, S.C. Stiros, Earthquake clustering in the Eastern Mediterranean during historical times. J. Geophys. Res., **101**, 6083–6097 (1996)
3. S.C. Stiros, P.A. Pirazzoli, J. Laborel, F. Laborel-Deguen, The 1953 earthquake in Cephalonia (Western Hellenic Arc): coastal uplift and halotectonic faulting. Geophys. J. Int., **117**, 834–849 (2007)

Chapter 10
Kinematic GPS River Profiles from Crete

In addition to the kinematic GPS profile of the river described in Chap. 5, three more profiles were carried out, and the results of these surveys are described here. The locations of these three rivers are plotted in Fig. 5.7 (Figs. 10.1, 10.2, 10.3, 10.4).

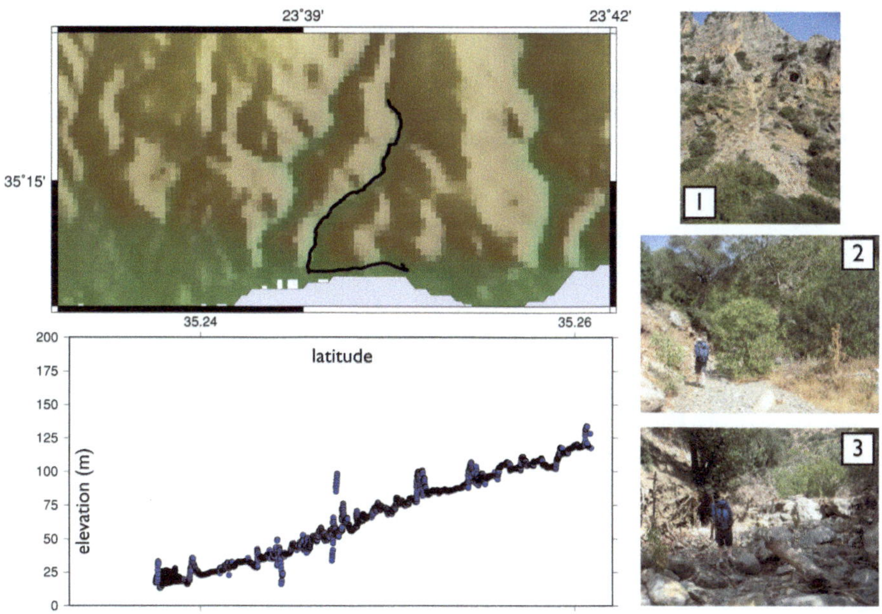

Fig. 10.1 The second river (Mouth at Agh. Ioannis), on the south coast of Crete, had a very open river valley and we were therefore able to produce a much more accurate river profile using kinematic GPS. *Photograph 1* was taken looking north–east from the highest point that we reached in the river profile and shows the upper course of a minor tributary. *Photographs 2* and *3* show the open character of the channel

B. Shaw, *Active Tectonics of the Hellenic Subduction Zone*, Springer Theses, 167
DOI: 10.1007/978-3-642-20804-1_10, © Springer-Verlag Berlin Heidelberg 2012

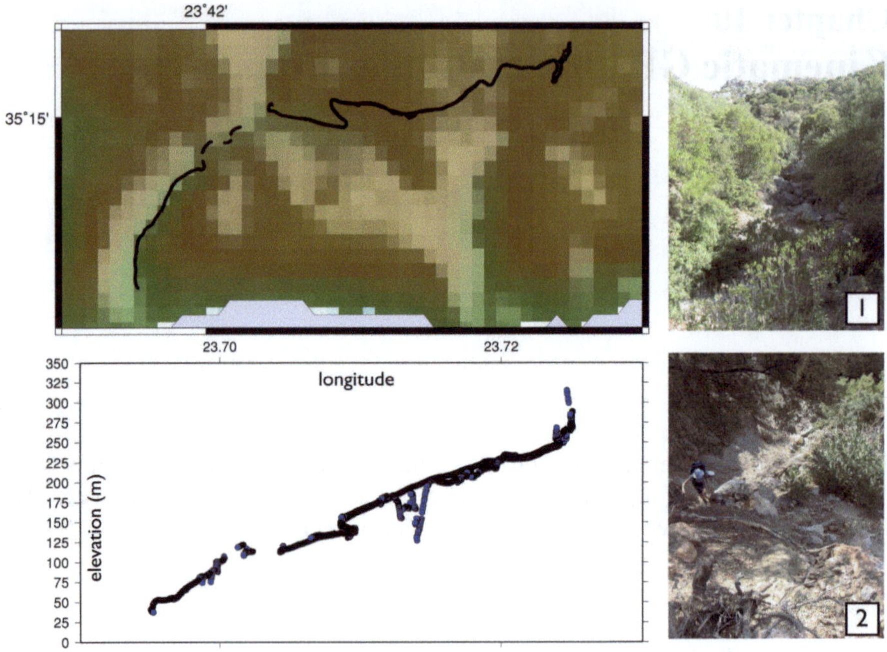

Fig. 10.2 Constraining the profile of river C, also on the south coast of Crete, was unsuccessful due to the extremely high density of vegetation cover. Clear nick-points could be recognised in the field (e.g. point 2 is a photograph taken looking vertically downward), but cannot be resolved in the resulting DEM. Parts of this profile were conducted along a road running alongside the river when the valley became impassable due to the density of vegetation

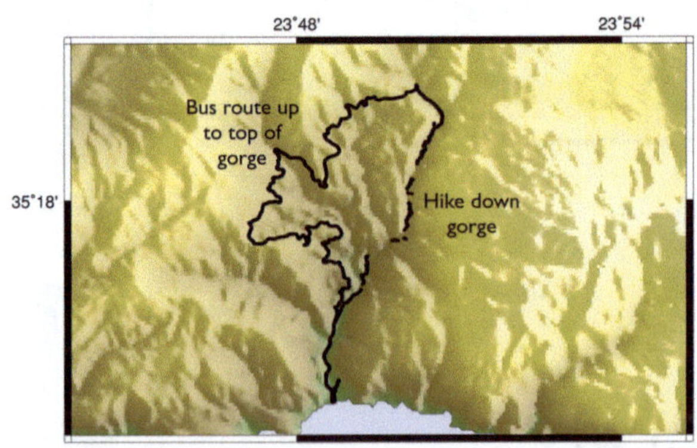

Fig. 10.3 Map view of the fourth river (Agh. Irini gorge). This is the longest river profile attempted, down a hiking trail running along the Aghia Irini gorge. GPS data was also collected from the front cab of the bus between sea-level and the top of the gorge for comparison with the SRTM and ASTER DEMs. This data is of excellent quality, as up to ten satellites were visible at any time. However within the river valley, severe data holes in the river profile occur where the channel is extremely narrow and deep, blocking the majority of the sky from view. In these areas, a maximum of three satellites could be seen at any time

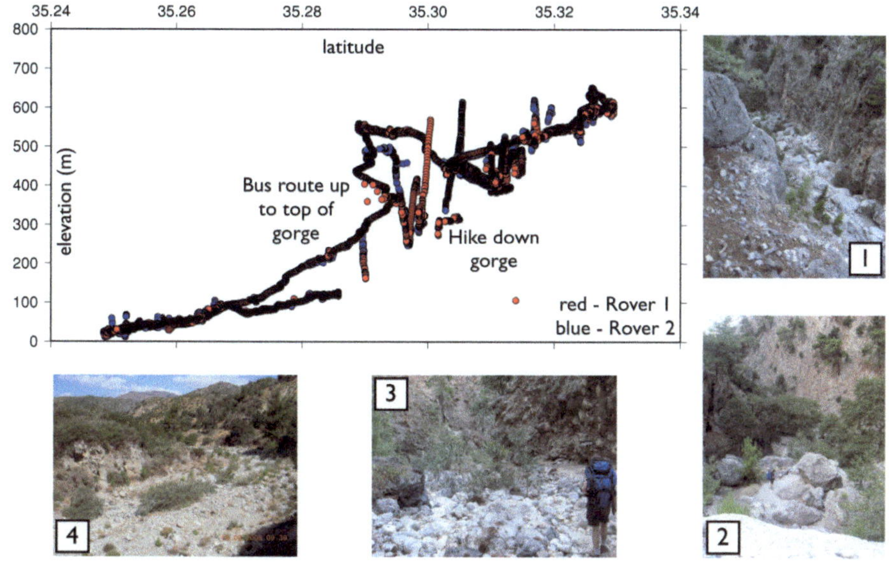

Fig. 10.4 The final attempt at mapping a river profile was carried out in the Aghia Irini gorge (river D). A popular hiking trail runs the length of the gorge, along the river channel. For this hike, we carried two roving receivers in order to collect as much data as possible. The *upper profile* shows the extremely accurate DEM of the bus route to the top of the gorge: the driver allowed us to sit in the front cabin, where between seven and ten satellites were visible at all times. In contrast, the DEM produced of the river channel itself is very poor, particularly in the area of interest, where nick-points and channel constrictions were observed (*Photographs 1–3*). The spectacular slot-canyon made the GPS survey highly impractical, as fewer than four satellites could be seen at any time. This method is therefore of limited utility in mapping river profiles in areas of dense vegetation, and in areas where river gorges are narrow and steep-sided. The approximate positions of nick-points can be recognised as the regions where data is particularly poor due to associated channel constrictions, but the resulting DEMs are of low quality. *Photograph 4* shows the broad, flat channel in the lower course of the river. Here the GPS data is excellent

MIX
Papier aus verantwortungsvollen Quellen
Paper from responsible sources
FSC® C105338

If you have any concerns about our products, you can contact us on
ProductSafety@springernature.com

In case Publisher is established outside the EU, the EU authorized representative is:
Springer Nature Customer Service Center GmbH
Europaplatz 3, 69115 Heidelberg, Germany

Printed by Libri Plureos GmbH
in Hamburg, Germany